THE
Pursuit of
Loneliness

THE
Pursuit of
Loneliness

American Culture at
the Breaking
Point

Philip Slater

With a New Introduction by
Todd Gitlin

Beacon Press   Boston

Beacon Press
25 Beacon Street
Boston, Massachusetts 02108-2800

Beacon Press books
are published under the auspices of
the Unitarian Universalist Association of Congregations.

97 96 95 94 93 92 91 90   1 2 3 4 5 6 7 8

Library of Congress Cataloging-in-Publication Data
Slater, Philip Elliot.
The pursuit of loneliness : American culture at the breaking point /
Philip Slater ; with an introduction by Todd Gitlin.—3rd Beacon
pbk. ed.
p. cm.
Includes bibliographical references.
ISBN 0-8070-4201-3
1. United States—Social conditions   1945–   2. National
characteristics, American.   3. Conflict of generations—United
States.   I. Title.
HN65.S565   1990
306′.0973′0904—dc20   89-43079

To Wendy, Scott, Stephanie, and Dashka

I'm empty and aching and I don't know why
Counting the cars on the New Jersey Turnpike
They've all come to look for America
                                    PAUL SIMON

Contents

Introduction

Todd Gitlin

Sometime in 1970, a friend of mine told me to read *The Pursuit of Loneliness*. My friend had taken time away from graduate study in sociology to work as a political organizer. In his midtwenties, he was no hippie—if anything, he was a craftsman of organization, an apostle of order, and an unabashed competitor. He was a family man and did not suffer communes gladly. He blamed anarchic looseness for having distracted the New Left from serious, rational political thinking. He believed in Puritan virtues that are today called traditional—hard work, family loyalty, individual responsibility. He took his politics straight. At least on alternate days I had a lot of sympathy with that position myself. So I was intrigued that my friend should recommend a book reputed to have a hippie edge—a book that said the American predicament lay deeper than politics or economics, deep in the normalcies of a self-destructive culture.

My friend said that once I read the parable that starts the book, I wouldn't be able to put the book down—in any sense. He was right. Philip Slater's self uglifying American, inventing and reinventing his apartness, building a prison with bars of freedom, was a character I could recognize. His individualism began with solitude and ended in paranoia—a logic that linked Walden with Fort Apache. This was the American journey into the Vietnam swamp. But it was also the flight to the suburbs—the ranch house fortified against intruders. Almost twenty years later, the parable still terrifies me.

*   *   *

At first glance, *The Pursuit of Loneliness* may seem sunk in a bygone time. Epigraphs from Dylan, Lennon and McCartney, Jagger and Richards, and the Grateful Dead mark it as a product of the 1960s—and aren't the sixties long gone, leaving behind nothing but nostalgia binges and ghostly traces in TV commercials for baby boomer gear? A stark prophetic tone stamps *The Pursuit of Loneliness* with the anguish and hope that were the very hydrogen and oxygen of the time when it was written—and aren't we all moderate now about everything but moderation?

But a book can have the truth of its exaggerations, and a prophecy is not to be judged only by the failure of its particular predictions. In reading *The Pursuit of Loneliness*, it is easy enough to spot the dated observations. Despite scientific pretensions, social science is always running afoul of history and its unintended consequences; as prophets, sociologists are generally no better than the economists who failed to anticipate, say, the stagflation of the seventies or the socially lopsided boom of the eighties. Who, in the fifties, anticipated the Vietnam war, psychedelic drugs, or student revolt? Who, in the sixties, anticipated Watergate, the oil crisis, and the social consequences of spiraling real estate prices? So it should not come as a surprise that a book which set out to explain the social and cultural upheaval known as "the sixties" to its own era should sound, at times, as if it comes from an exploded star. What is remarkable is how much of this book remains distressingly relevant to the present.

The Toilet Assumption, for one—the belief that social unpleasantness, once flushed out of sight, ceases to exist—remains central to American culture. Examples are many. Since the North American continent afforded the luxury of space, escape and evasion became national pastimes. The energy crisis of the seventies was supposed to have been "fixed" by improved plumbing, in which Alaskan oil—out of sight, out of mind—featured prominently. When the *Exxon Valdez* ran aground and spilled ten million gallons of crude oil into Prince William Sound, attention focused on the pathetic figure of the errant captain. There were justifiable demands that Exxon clean up the beaches. Perhaps this fit the Janitorial Corollary: whatever spills, someone will clean up without the rest of us having to trouble ourselves about it. There is resistance to the systemic conclusion that dependence on fossil fuels will poison the world we live in (quaintly known as "the environment" as if it were something we could keep at arm's length).

To take another example, the wide open spaces of suburbia were meant to be escapes from urban congestion. They had the effect of rendering the poor, as Michael Harrington wrote in the early sixties, "invisible" to the middle classes. The embarassing and infuriating thing about the homeless of subsequent decades is that they have the bad taste to foist their misery on the rest of society. They have violated the implicit contract according to which they submit to sequestration among their own kind. "Reform" would consist of some way of moving the homeless to "shelters" where they would keep out of our way. Meanwhile, we do not question the social machinery which has produced homelessness. There is no public outcry against the tax abatements which enrich the "developers" (they are also destroyers) who dump the poor over the side as a drag on upscale growth. We want someone to "take care of the problem," meaning get it out of our faces. It remains true, likewise, as Philip Slater wrote, that "our approach to social issues inevitably falls back on a cinematic tradition in which social programs are resolved by a gesture." Think of the gestures of the eighties: "Just Say No"; the death penalty; Star Wars; "tax simplification"; "enterprise zones"; cold fusion.

Americans are not, however, uniquely evasive—the Italian romance with the automobile easily rivals the American—and I am still powerfully unconvinced that our evasiveness is rooted in a sort of "natural selection" for what might be called the runaway character. A better explanation might be that American culture, lacking aristocratic or time-honored understandings of the good life, perpetually has to scramble for standards of conduct. Moral crusades are one outlet; another is the abstraction of money ("the bottom line"); a third is an overvalued individualism—the material *and spiritual* acquisitiveness at which Slater directs his most scathing lines.

"Individualism," Slater writes, "is rooted in the attempt to deny the reality of human interdependence." Details change, but the substratum of his argument remains powerful. This is an old American story, of course: it is striking how many of our national darlings, from Lincoln to Michael Jackson, from Tom Sawyer to Jerry Rubin to Reagan and Rambo, are self-invented. (F. Scott

Fitzgerald was never more wrong than when he wrote that American lives have no second acts; it is closer to the truth that they have nothing but second acts.) We are constantly told, and tell ourselves, that we, unique and free-standing individuals, should "be all that we can be." Yet the unacknowledged testing ground and presupposition of that ambition is the army, the art and celebrity markets, or some other social institution. The point is that the absolutes of individualism are impossible, that personal achievement denies a dependency it takes for granted.

In pointing to the foundations and consequences of American individualism, Slater is an unabashed moralist, of course. He condemns individualism for getting out of control and driving cooperation into a furtive, underground existence. But he is not looking for a quick fix of his own. He is sophisticated both psychoanalytically and sociologically, and he understands that individualism does not simply invent itself; it is part of an intricate social complex. His argument, very roughly, is that an excess of individualism thrives in the isolated family, where inordinate Oedipal pressure generates extreme narcissistic hunger, inner futility, stressful self-making, ferocious and disabling competition, and a "servility toward technology" which aims to solve the problems generated by individualism and rootlessness in the first place. At the same time, unbridled individualists try to compensate for their centrifugal tendency by delegating authority upward. Unable to find social solidarity horizontally, through cooperation, they search for it vertically, through authoritarian imposition. Much of Americans' vaunted individualism is a fevered denial of our conformity and a longing for guidance.

Our vehement insistence, in chorus, on our uniqueness conceals an even uglier side of this individualism—terror that somebody out there is scheming against us, keeping us from getting our due. Individualism, Slater argues, is the root of the military and economic and cultural mobilization which has been institutionalized since 1947 in the national security state. As that expert on American culture, Lyndon Johnson, said in 1966: "There are three billion people in the world and we only have two hundred million of them. We are outnumbered fifteen to one. If might did make right they would sweep over the United States and take what we have. We have what they want." In analyzing our national isolationism, Slater was the first American sociologist to develop the sixties' radical idea that our domestic arrangements and foreign policy are the inside and outside of the same phenomenon. (Later this point would become a commonplace of feminist theory.) Audaciously, Slater tries to show that the misunderstanding which is American individualism has terrible consequences for the American sense of mission in the world. It gives rise, in short, to paranoia. How well does his argument work?

It is not fashionable today to ask how the American (or any other) way of life might be implicated in the violence which is at the disposal of the state. Indeed, in his 1989 inaugural address, President George Bush declared that the lesson to be drawn from the Vietnam war was that we ought no longer to be debating that lesson. He wishes, in other words, to make foreign policy unencumbered. The Toilet Assumption reigns—off the nightly news, out of mind. Meanwhile, in post-liberal quarters, we still hear echoes of the cavalier claim that the war was a regrettable mistake—as if the blasting of so much life and treasure in the longest war in the history of the nation could be attributed to nothing more than faulty calculation. And so it is useful to think again about what that atrocious war had to do with the political culture that prosecuted it and refused, for so long, to let go.

Slater's idea is a daring and ingenious one. He calls attention to the bureaucratic nature of a war whose characteristic form of violence was "killing from a distance." The heretical question—heretical even for the left, which believes that "the people" are innocent of their rulers' military expeditions—is: What might this tell us about the war's psychic pleasures? Slater argues that middle-class Americans feel helpless to identify the social forces that damage their lives. Helplessness leads to indifference. Because we are hell-bent on autonomy, we don't know who to blame for our lingering feeling that life is beyond our control. "Our fondness for violence at a distance," Slater writes, "is both an expression of, and a revenge against, this indifference. We sent bombers to destroy 'communism' in Vietnam to avoid meeting our needs for cooperation at home." Leave aside the quotation marks that suggest that the social system America was trying to pulverize in Vietnam was not really communism. Communism it was—although that fact, and the terrible repression that followed the war, does not begin to explain the ferocity of America's technological rage, or begin to justify it. The problem remains: What was distinctive about this war? Slater's answer is that in Vietnam we tried to prove we were free by projecting our own subjection onto the other side of the world ("the next thing you know, the communists will be coming ashore in California"), and turning brute technology loose as our instrument of revenge. We wouldn't have to fear technology if we could use it to scorch somebody else's earth, far away across the ocean.

An extraordinary idea, not to be dismissed out of hand. Slater is absolutely right to call attention to the fact that the distinctive thing about the Vietnam war was air power. The number of Americans who died in Vietnam was "only" four thousand greater than the number who died in Korea; but *between 1965 and 1973 the government of the United States dropped on Indochina, including Laos and Cambodia, more than four times the bomb tonnage it dropped on all of Europe and Asia during World War II.*

But Americans did not unanimously accept this slaughter. For one thing, after the Tet offensive in the winter of 1968, Americans were far less enthusiastic about the war. As soon as the war looked like a loss, the political class and public opinion began to look for a way out. Having found one, thanks to Watergate, they have refused to mobilize the armed forces since. The antiwar movement, for all its failings, was the most widespread and arguably the most effective antiwar movement in the history of the world; and it left its trace on American culture and politics, so that even the Reagan administration was willing to dispatch American troops only on weekend wars. Because Congress had finally placed limits on White House power, paranoia had to retreat underground—as in the secret funding of the contra war. The Iran-contra scandal, in other words, was the backhanded tribute that the White House paid to democracy. Democracy, a lumbering latecomer to the making of foreign policy, has had its effect.

Perhaps, in the end, indifference and violence are indeed related in the way Slater argues. Perhaps, since the Vietnam war,

American society has done less violence outside its borders because, in some respects, Americans have become more active, less fatalistic. One example: For all the cynicism and withdrawal of the last decade, today we take demonstrations for granted. For all the privatism, there is a countertendency—an unacknowledged, unphotogenic surge toward what the New Left called participation in the decisions that affect our lives, from work circles and management teams to movements against nuclear power and toxic waste dumps. I live in a city which declares itself a nuclear free zone (the navy doesn't agree), along with several score others around the country. What happened after the sixties was not the millennium but not the fifties either.

Two decades of history will embarrass a great many social observations. If it is striking how much of Philip Slater's analysis still repays attention, it is equally striking how much it carries the apocalyptic temper of its time. Consider the Manichaean excess, the naively American notion that "having lately become in many ways the worst of societies we could just as easily become the best." Consider the edgy subtitle the book's earlier two editions carried: "American Culture at the Breaking Point." But American culture didn't break—it bent, twisted, enfolded, decayed, and otherwise went on. In a certain respect, the result is more disconcerting than the millennial expectation that a new culture would overtake the old. It is the old story of the frog in the pot who would hop out of boiling water but, if the temperature is turned up a degree at a time, stays to its death.

*   *   *

Today the most oft-repeated critique of American culture is that we are too slovenly to compete successfully with the Japanese. Uneasy about individualism, Americans are still not willing to give up its excesses—or automobiles. The love side of the counterculture, predicated on an exaltation of youth, could not outlast the aging of the baby boom cohorts or the economic anxiety attendent upon the fading of the American Century. Which is not to say that Americans are content with their institutions. There is criticism galore, though much of it has been narrowed and specialized. The left, environmentalism, feminism, "New Age"—each has its more or less private language and its self-enclosed constituency. If you belong to the academy, it is all you can do to "keep up" with your ever-narrower specialty. An obsession with method, mirroring the culture's obsession with "information," buries ideas. If you resist specialization, you are drawn toward the hall of mirrors called "theory"—it becomes a full-time occupation to find your way down the shimmering corridors. Odds are that you do not take pleasure from what you read, for social criticism today is professionalized, self-enclosed, and segregated, like the rest of our intellectual discourse. To write accessibly, on the other hand, is to take seriously the democratic faith. It is to render tribute to the powers and curiosity of that crucial figure, the "general reader." Possibly this quaint-sounding creature is real, possibly chimerical. In any case, it is an indispensable act of faith to address her. *The Pursuit of Loneliness* is still worth reading because it honors the mission of sociology not to train a caste of knowers but to contribute to society's knowledge of itself.

Preface to the 1976 Edition

My reasons for writing a revised edition of *The Pursuit of Loneliness* are twofold. The first is that parts of it seem to me dated. This is hard to avoid in a contemporary social critique. The book was very much a product of the sixties—imbued with the spirit, turmoil, and excitement of those times. Yet the bulk of what I wrote about dealt with chronic sources of stress in our society— I felt a need to present these ideas in a context that wouldn't be subject to changing social moods.

I realize that this is an impossible dream. Our society is changing very fast and almost anything could happen in the next five years. It's even possible that another era of youthful turmoil and confrontation could arise. Yet one thing I'm certain of: if it does, it will be different. I suspect it will be less luxurious, less middle-class, more tied to bread-and-butter issues. So although this edition will no doubt become quickly dated, too, the effort seems worth making.

The second reason for revision is stylistic. I fondly believed that I had risen above the obscurities of academic prose in the first edition, but I've had some feedback to the contrary. While the beast is by no means dead even yet, I've done my best to make the book simpler and clearer.

I'm sure there will be differences of opinion about what I chose to leave out and what I've added. For the most part the passages eliminated dealt with university struggles, generational conflict, the "counterculture," and political confrontations. Not that much isn't to be learned from studying these things, but much of the material was posed in a way that no longer seems particularly relevant or gripping. At times I've felt that a point I wanted to make required some reference to the sixties, but much of that material has been dropped.

A lot of the material on the Vietnam War, on the other hand, was retained, since, although the war has ended, violence and the technological impulse are still very much with us. There's still a lot to be learned from that painful episode in our history.

It's already apparent that some people's favorite parts of the book have been excised, but my goal was to write a new version, not a longer one. Those who feel intense nostalgia for or interest in the sixties will probably want to stay with the old edition.

As to additions, there's no particular pattern. I've put them in everywhere, for I've had new thoughts on almost all aspects of the original book. The final chapter, furthermore, dealing with economic issues, is entirely new, since they seem particularly acute at the time of this writing, and subject to an unusual amount of official obfuscation.

I've had mixed responses to my handling of gender pronouns. Some think it silly and "inconvenient" to deviate from the impersonal "he." But change is always inconvenient; and through such trivial innovations you learn who the society is arranged for and why. The "convenience" in this case is for men, and women are quite right in objecting to a practice that daily reinforces the thought patterns that enslave the mind and sap the will.

Preface

Once upon a time there was a man who sought escape from the prattle of his neighbors and went to live alone in a hut he had found in the forest. At first he was content, but a bitter winter led him to cut down the trees around his hut for firewood. The next summer he was hot and uncomfortable because his hut had no shade, and he complained bitterly of the harshness of the elements.

He made a little garden and kept some chickens, but rabbits were attracted by the food in the garden and ate much of it. The man went into the forest and trapped a fox, which he tamed and taught to catch rabbits. But the fox ate up the man's chickens as well. The man shot the fox and cursed the perfidy of the creatures of the wild.

The man always threw his refuse on the floor of his hut and soon it swarmed with vermin. He then built an ingenious system of hooks and pulleys so that everything in the hut could be suspended from the ceiling. But the strain was too much for the flimsy hut and it soon collapsed. The man grumbled about the inferior construction of the hut and built himself a new one.

One day he boasted to a relative in his old village about the peaceful beauty and plentiful game surrounding his forest home. The relative was impressed and reported back to his neighbors, who began to use the area for picnics and hunting excursions. The man was upset by this and cursed the intrusiveness of human beings. He began posting signs, setting traps, and shooting at those who came near his dwelling. In revenge groups of boys would come at night from time to time to frighten him and steal things. The man took to sleeping every night in a chair by the window with a loaded shotgun across his knees. One night he turned in his sleep and shot off his foot. The villagers were saddened by this misfortune and thereafter stayed away from his part of the forest. The man became lonely and cursed the unfriendliness of his former neighbors And all these troubles the man saw as coming from outside himself, for which reason, and because of his technical skills, the villagers called him the American.

My purpose in writing this book is to reach some understanding of the forces that seem to be unraveling our society. I want to talk less about what happens to people than about what people do—to themselves, to each other. Hence I'm writing primarily about those whose behavior has the greatest impact on the society. Most of what I have to say is about middle-class life, which should be kept in mind whenever it begins to sound as if all Americans attend college or own their own homes. Finally, I'm writing for and about Americans. This doesn't mean the book is relevant only for Americans, but the problems discussed are most fully developed in America, and it's in America that the major battles will be fought.

*   *   *

A traveler returning to her own country after spending some time abroad receives a fresh vision of it. She still wears her traveler's antennae—a sensitivity to nuances of custom and attitude that helps her adapt and make her way in strange settings.

Re-entering America, one is struck by the grim monotony of American facial expressions—hard, surly, and bitter—and by the aura of deprivation that informs them. One goes abroad fore-warned against exploitation by grasping foreigners, but nothing is done to prepare the returning traveler for the fanatical acquisitiveness of her compatriots. It's hard to get reaccustomed to seeing people already weighted down with possessions acting as if every object they didn't own were bread withheld from a hungry mouth.

These impressions are heightened by the contrast between the sullen faces of real people and the vision of happiness television offers: men and women ecstatically engaged in stereotyped symbols of fun—running through fields, strolling on beaches, dancing and singing. Smiling faces with chronically

open mouths express their gratification with the bounties offered
by our culture. One begins to sense a wide gap between the
fantasies Americans live by and the realities they live *in*. Americans know from an early age how they're supposed to look when
happy and what they're supposed to do or buy to be happy. But
for some reason their fantasies are unrealizable and leave them
disappointed and embittered.

The traveler's antennae disappear after a time. The impressions fade, and re-entry is gradually completed. America once
again seems familiar, comfortable, ordinary. Yet some uneasiness
lingers on, for the society seems troubled and self-
preoccupied—as if large numbers of Americans were beginning
to scrutinize their own society with the doubtful eyes of a traveler.

                                                    PHILIP SLATER

THE
Pursuit of
Loneliness

ONE

I Only Work Here

He said his name was Columbus,
And I just said, "good luck."
DYLAN

All the lonely people—
Where do they all come from?
LENNON & MCCARTNEY

One of the first goals of a society is to make its inhabitants feel
safe. More of our collective resources are devoted to national
security and local safety than to any other need. Yet Americans
feel far less safe, both at home and abroad, than they did fifty
years ago. Our nuclear arsenal, the guns under pillows, and the
multiple lock on city doors betray our fears without easing them.

Cold warriors have always attributed this uneasiness to the
growth of communism—its military might, its domestic seductiveness, its success in winning the allegiance of third world
peoples. Others see a decline in moral values. But before any
political conversation goes very far, it becomes clear that everyone thinks something is wrong with American society. Sometimes a particular group or class is blamed, sometimes our leaders, sometimes our political or social institutions. But most
people seem to agree that the machinery isn't functioning as it
should.

The great defect in all these accusations is that the blame is
always placed elsewhere. Who, then, fosters all this misery, decay, and disruption? Even when institutions are blamed it's with
a curious detachment, as if they had nothing to do with us as

people—as if we didn't support them with our own wishes, motives, and actions.

One assumption underlying this book is that every morning
all 200 million of us get out of bed and put a lot of energy into
creating and re-creating the social calamities that oppress, infu-
riate, and exhaust us.

People and Societies

Our society is much divided. There are fewer and fewer things
that people agree about. The furor of the sixties has died down,
but the conflicts that caused it persist. The bitterness has merely
been diluted with despair.

The war is over, but peace has hardly reached epidemic pro-
portions. Our economic structure is crumbling. Cities are going
bankrupt. We are exhausting our physical resources while our
human ones lie fallow. The quality of all our services—transportation, communication, health, education, and so on—deterio-
rates, while their costs increase. Some people can't afford to heat
their homes because we all want to ride expensive vehicles on
crowded roads at high speeds, killing one another and polluting
the atmosphere. During the years that we tried heroically to an-
nihilate the Vietnamese population and destroy its countryside,
we slaughtered half a million Americans and did massive damage
to our own landscape just driving around. Intent seems to have
nothing to do with it.

During the anticommunist hysteria of the early fifties a survey
respondent remarked that "so many people in America are eager
. . . to fall into the trap set by Communist propaganda."[1] The
poor soul's fears proved exaggerated, but the language was revealing. Why "eager"? Even 20 years ago, it seems, American
society had its flaws.

The idea that people have a deep attraction to what is alien to
their society expresses a simple truth: that all societies frustrate
certain human needs and satiate others—that the fit between
humanity and any particular society's idea of what humanity
should be is never very exact.

The emotional repertory of human beings is simple and
straightforward. We're built to feel warm and happy when we're

caressed, angry when frustrated, frightened when threatened, hurt when rejected, offended when insulted, sad when abandoned, jealous when excluded, and so on. But every culture holds some of these human reactions to be unacceptable and tries to warp its participants into some peculiar limitation. Some learn not to laugh, others not to cry; some not to love and some not to hate. Since humans are rather pliable within limits, this warping is successful, but it exacts a heavy toll in misery, neurosis, and physical illness.

Sex Roles: An Example

The masculine ideal in our culture, for example, has traditionally been one of almost complete emotional constipation. Anger might be encouraged in certain settings, but tears or joyous effusions never. Men could choose between a mask of mute stolidity and one of pompous intellectuality. Not that this tradition has disappeared—most men even today are stuck with this choice between articulate and inarticulate zombiehood. Vulnerability and spontaneity are still taboo for men although the ideal is now being questioned. Women, on the other hand, have traditionally been encouraged to express all feelings *except* anger, and have aroused intense anxiety in males unless they muted their assertiveness and disguised their power.

In the Victorian era women were supposed to mute their sexuality as well. Even when sexuality for women came back in vogue a sharp constraint was placed upon it: women were to be sexual only in ways that were nonthreatening to men. They were to be sexually aroused only in response to a man and on his initiative, and their sexuality was to be clothed in the garb of submission and docility. The idealized sex symbols were, and still are, unassertive, weak-willed, and stupid.

These rules are deeply ingrained and express themselves in body postures. Men in our culture tend to have masklike, impassive faces, hard jaws and tight mouths, flat, constricted voices, and rigid unresponsive bodies—good for doing, poor for experiencing. Women, on the other hand, tend to hold themselves in white-flag attitudes that chronically reassure men that they come in peace: smiles, head to one side, and nonthreatening hands.

The proper way for a woman's hands to look in our culture is helpless and ineffectual. This is done largely by treating them as if they were broken at the wrist. It's very hard for American women in spontaneous conversation with men to let their hands form a straight line from wrist to fingertips. The hands are dropped, flipped, waved, or turned back against the body. What they do *not* do is point. A woman who allows herself to point her finger directly at a man is usually called a ballbreaker. If a woman observes these body rules, on the other hand, she will ordinarily be accepted by men whatever her political views on sex roles.

What does a person do with feelings that are disapproved by his or her society? To some extent they can be released in the guise of some other feeling that carries no disgrace. In our society, for example, women often dissolve in tears instead of raging in anger, while men get angry when their feelings are hurt and they feel like crying.

Another solution is through vicarious expression. Women can get a little satisfaction by getting men to express anger for them, just as men can get a little satisfaction by getting women to cry for them. This is why women have traditionally placed such a high value on aggressiveness in men, and why men have put so much stress on responsiveness in women. Women are trained to feel ashamed of their own assertiveness, and hence feel that the least a man can do is do it for them. Women who browbeat passive husbands often complain that all they want is a man who can stand up to them, and there's a great deal of truth in this. If they're made to feel bad about their *own* aggressiveness, they can't be blamed for being harsh with the person who's supposed to be doing it for them. On the other hand, if they could accept their own assertiveness as a human right, without guilt, they wouldn't need to demand it from a man.

Similarly, men insist that women be emotionally responsive and pliable and accuse them of being unfeminine when they're not. This is because men themselves tend to be ashamed of soft or yielding behavior and insist that women do it for them. If they were comfortable being less rigid and constipated emotionally, they wouldn't need to make this demand on women.

All these roles and patterns are under attack now by the women's movement and other groups, but no matter how much the roles change—even if they were to reverse, in fact—the problem remains the same: every society, community, or group tends to look down on *some* human emotion, and to discourage its expression by certain categories of people.

New No-No's

In American society, for example, we have been liberated from the cruder forms of sexual inhibition, although most Americans are still uncomfortable with any eroticism that doesn't lead to orgasm. Furthermore, some groups in our society explicitly wave the banner of emotional expression: we should not hold back feelings, we should be "up front" with them. But this by no means leads to an "equal opportunity law" among feelings. For these groups there are new pariahs: jealousy and guilt. Since "jealousy is merely the product of a capitalistic upbringing" and reveals a "possessive attitude" toward the loved one, one is not supposed to feel it even under the most extreme circumstances. Guilt has also been drummed out of the corps as betraying an insufficient ability to live in the here-and-now. Young Americans struggling earnestly to ignore the moralities they grew up with, or soberly plodding through the boggy moors of open marriages and triangular relationships, have a rather touching faith in the plasticity of human emotion. I often find myself torn between tears of admiration at their attempts to break new ground, and waves of hilarity at the way they repeatedly fall into holes they pretend aren't there—a bit of slapstick worthy of Mack Sennett. Their bewilderment and guilt when they feel things they're not supposed to is exactly what Victorian women felt when they were sexually aroused, or what Quakers feel in the grip of rage. (Although it may be the first time people have felt guilty for feeling guilty.)

Some people, of course, seem able to bring it off. As Margaret Mead has often pointed out, the demands that cultures lay on their participants are harder on some temperaments than on others. It helps if you happen to be a cold fish in a Puritanical society, or a phlegmatic turtle in a Quaker one; and a shallow, fickle con artist is in a similarly enviable status in counter-culture circles today, although it's poor form to mention it.

But whatever the feeling that's tabooed, its suppression causes distress. A certain amount of roundabout relief may be gained vicariously or through discharging other feelings, but some discomfort always remains. Usually it takes physical form in chronic muscular tension. Such tension and its secondary effects cause much illness, perhaps most of it, and a lot of the decay we attribute to "old age" is just the erosion caused by each society trying to cram humans into some emotional mold that doesn't quite fit.

The Seeds of Change

This distress is the origin of some of the "eagerness" to abandon the values of one's own society. There's no use pointing out that most, if not all, societies impose some sort of inhuman stress on their participants—if a man has a toothache it isn't helpful to tell him that tooth decay is widespread. A society is like a dwelling: each one has annoying limitations, defects, and inconveniences, but a particular irritation may become insupportable and provoke the occupant to move. Even if new annoyances are acquired in the process, they're at least novel. Furthermore, the stresses in some societies are much more severe than in others. If a lot of people complain about the quality of life in a society, it seems reasonable to assume that something is wrong with it.

It's important to remember that although a society may try to squelch some human attribute, it can never completely succeed. There is always some devious channel through which the disapproved trait can find at least partial release. Thus although the Germans have always stressed order, precision, and obedience to authority, they have periodically exploded into revolutionary chaos and are much influenced by romantic Gotterdammerung fantasies. In the same way, there is a cooperative underside to competitive America, a rich spoofing tradition in ceremonious England, an elaborate pornography accompanying Victorian prudishness, and so on. Rather than saying Germans are obedient or Anglo-Saxons stuffy or puritanical, we might more correctly say that Germans are preoccupied with the problem of authority and order, Anglo-Saxons, with the control of emotional and sexual expression. The issues about which people in any society feel strongly are based in a conflict. One side is always talked up, the other shouted down, but never quite successfully.

These opposing forces are much more equally balanced than the society's participants like to recognize—otherwise there would be no need for suppression. Life would be much less frantic if we were all able to recognize the diversity within ourselves, and could abandon our futile efforts to present monolithic self-portraits to the world. Probably some exaggeration of our internal homogeneity is necessary, however, in order for us to act with enough consistency to permit smooth social relations.

An individual who "converts" from one viewpoint to its exact opposite appears to himself and others to have made a gross change, but actually it involves only a very small shift in the balance of a persistent conflict. Just as only one percent of the voting population is needed to reverse the results of an American election, so only one percent of an individual's internal "constituencies" need shift in order to transform him from voluptuary to ascetic, from policeman to criminal, from rich honor student to radical terrorist, from Communist to anti-Communist, or whatever. The opposite sides are as evenly matched as before, and the apparent change merely represents the desperate efforts made by the new internal "majority" to consolidate its shaky position of dominance. The individual must expend just as much energy shouting down the new "minority" as he did the old. Some of the most dedicated witch hunters of the 1950s, for example, were ex-Communists.

*   *   *

Societies are more complex than individuals, and have more available outlets for unacceptable impulses. Still, the pressures remain, and often build up to a pitch of intense distress. Whether today's pressures could lead to a cultural flip-flop or not I cannot say, but I would like to suggest three human desires that are deeply and uniquely frustrated by American culture:

(1) The desire for *community*—the wish to live in trust, cooperation, and friendship with those around one.

(2) The desire for *engagement*—the wish to come directly to grips with one's social and physical environment.

(3) The desire for *dependence*—the wish to share responsibility for the control of one's impulses and the direction of one's life.

When I say that these three desires are frustrated by American culture I don't want to conjure up romantic images of the individual struggling against society. In each case we participate eagerly in producing the frustration we endure—it isn't merely something done to us. For each of these desires is subordinate to its opposite in the American character. The thesis of this chapter is that Americans have voluntarily created, and voluntarily maintain, a society that increasingly frustrates and aggravates these secondary yearnings to the point where they threaten to become primary. Groups that in any way personify this threat have always been feared in an exaggerated way, and always will be until Americans are able to recognize and accept those needs within themselves.

I. COMMUNITY AND COMPETITION: GETTING TOGETHER

We are so used to living in an individualistic society that we need to be reminded that collectivism has been the more usual lot of humans. Most people in most societies have lived and died in stable communities that took for granted the subordination of the individual to the welfare of the group. The aggrandizement of the individual at the expense of his neighbors was simply a crime.

This is not to say that competition is an American invention— all societies involve some mixture of cooperative and competitive institutions. But our society lies near the competitive extreme, and although it contains cooperative institutions, we suffer from their weakness and peripherality. Studies of business executives reveal a deep hunger for an atmosphere of trust and fraternity with their colleagues. The competitive life is a lonely one and its satisfactions short-lived, for each race leads only to a new one.

In the past our society had many oases in which one could take refuge from the frenzied invidiousness of our economic system—institutions such as the extended family and the stable local neighborhood in which people could take pleasure from something other than winning symbolic victories over their neighbors. But these have disappeared one by one, leaving us more and more in a situation in which we must try to satisfy our vanity and our needs for intimacy in the same place and at the same time. This has made the appeal of cooperative living more seductive, and the need to suppress our longing for it more acute.

The main vehicle for the expression of this longing has been the mass media. Popular songs and film comedies for fifty years have been engaged in a sentimental rejection of our dominant mores, maintaining that the best things in life are free, that love is more important than success, that keeping up with the Joneses is futile, that personal integrity should take precedence over winning, and so on. But these protestations must be understood for what they are: a safety valve. The same man who chuckles and sentimentalizes over a happy-go-lucky hero in a film would view his real-life counterpart as frivolous and irresponsible, and suburbanites who philosophized over the back fence with complete sincerity about their "dog-eat-dog-world," and what-is-it-all-for, and you-can't-take-it-with-you, and success-doesn't-make-you-happy-it-just-gives-you-ulcers-and-a-heart-condition, were enraged in the sixties when their children began to pay serious attention to these ideas. To the young this seemed hypocritical, but if adults didn't feel these things they wouldn't have had to fight them so vigorously. The exaggerated hostility that young people aroused in the "flower child" era argues that the life they led was highly seductive to middle-aged Americans.

When a value is strongly held, as individualism is in America, the illnesses it produces tend to be treated in the same way an alcoholic treats a hangover or a drug addict his withdrawal symptoms. Technological change, mobility, and individualistic ways of thinking all rupture the bonds that tie a man to a family, a community, a kinship network, a geographical location—bonds that give him a comfortable sense of himself. As this sense of himself erodes, he seeks ways of affirming it. Yet his efforts accelerate the very erosion he seeks to halt.

This loss of a sense of oneself, a sense of one's place in the scheme of things, produces a jungle of competing egos, each trying to *create* a place. Huge corporations are fueled on this energy—the stockholders trying to buy place with wealth, executives trying to grasp it through power and prestige, public relations departments and advertisers trying to persuade people that the corporation can confer a sense of place to those who believe in it or buy its products.

Americans love bigness, mostly because they feel so small. They feel small because they're unconnected, without a place. They try to overcome that smallness by associating themselves with bigness—big projects, big organizations, big government, mass markets, mass media, "nationwide, worldwide." But it's that very same bigness that rips away their sense of connectedness and place and makes them feel small. A vicious circle.

Notice the names of corporations: "Universal," "Continental," "International," "General," "National," "Trans-World"—the spirit of grandiosity and ego-inflation pervades our economic life. Corporations exist not to feed or supply the people, but to appease their own hungry egos. Advertising pays scant attention to price or quality and leans heavily on our needs for acceptance and respect. The economic structure of our society continually frustrates those needs, creating an artificial scarcity that in turn motivates the entire economy. This is why the quality of life in America is so unsatisfying. Since our economy is built on inflated vanity, rather than being grounded in the real material needs of the people, it must eventually collapse, when these illusions can no longer be maintained.

Much of the unpleasantness, abrasiveness, and costliness of American life comes from the fact that we're always dealing with strangers. This is what bureaucracy is: a mechanism for carrying on transactions between strangers. Who would need all those offices, all that paperwork, all those lawyers, contracts, rules and regulations, if all economic transactions took place between lifelong neighbors? A huge and tedious machinery has evolved to cope with the fact that we prefer to carry on our activities among strangers. The preference is justified, as are most of the sicknesses in American society, by the alleged economic benefits of bigness, but like many economic arguments, it's a con.

On the surface, it seems convincing. Any big company can undersell a little one. Corporations keep getting bigger and bigger and fewer and fewer. Doesn't that prove it? Survival of the fittest? Yet for some reason, what should be providing economic benefits to the consumer has in fact produced nothing but chronic inflation. If bigness lowers the cost of production, why does everything cost more and break sooner? Management, of course, blames it on labor, and each industry cites the rising prices of its own suppliers. Isn't it obvious that a few big nationwide companies can produce things cheaper than many local ones?

It all depends on what you leave out of your analysis (which is why a chimp pressing buttons randomly could predict as well as our economic forecasters). The fewer the companies, the less influence supply and demand have on prices. A heavy investment in advertising and public relations is necessary to keep a national reputation alive. And what about the transportation costs involved when all firms are national? Not to mention the air pollution costs, which are also passed on to the consumer. Chronic inflation suggests that someone is leaving something vital out of his analysis. How does one measure in dollars the cost of economic mistrust? It may be subtle, but it's clearly enormous.

The Great Illusion

It's easy to produce examples of the many ways in which Americans try to minimize, circumvent, or deny the interdependence upon which all human societies are based. We seek a private house, a private means of transportation, a private garden, a private laundry, self-service stores, and do-it-yourself skills of every kind. An enormous technology seems to have set itself the task of making it unnecessary for one human being ever to ask anything of another in the course of going about his or her daily business. Even within the family Americans are unique in their feeling that each member should have a separate room, and even a separate telephone, television, and car, when economically possible. We seek more and more privacy, and feel more and more alienated and lonely when we get it. And what accidental contacts we do have seem more intrusive, not only because they're unsought, but because they're not connected with any familiar pattern of interdependence.

Most important, our encounters with others tend increasingly to be competitive as we search for more privacy. We less and less often meet our fellow humans to share and exchange, and more and more often encounter them as an impediment or a nuisance: making the highway crowded when we're rushing somewhere, cluttering and littering the beach or park or wood, pushing in front of us at the supermarket, taking the last parking place, polluting our air and water, building a highway through our house, blocking our view, and so on. Because we've cut off so much communication with each other we keep bumping into each other, so that a higher and higher percentage of our interpersonal contacts are abrasive.

We seem unable to foresee that the gratification of a wish might turn out to be a monkey's paw if the wish were shared by many others. We cheer the new road that shaves ten minutes off the drive to our country retreat but ultimately transforms it into a crowded resort and increases both the traffic and the time. We're continually surprised to find, when we want something, that thousands or millions of others want it, too—that other human beings get hot in summer and cold in winter. The worst traffic jams occur when a mass of vacationing tourists start home early to "beat the traffic." We're too enamored of the individualistic fantasy that everyone is, or should be, different—that a man could somehow build his entire life around some single eccentricity without boring himself and everyone else to death. We all have our quirks, which provide surface variety, but aside from this, human beings have little basis for their persistent claim that they are not all members of the same species.

The Freedom Fix

Since our contacts with others are increasingly competitive, unanticipated, and abrasive, we seek still more apartness and thus accelerate the trend. The desire to be somehow special sparks an even more competitive quest for progressively more rare and expensive symbols—a quest that is ultimately futile since it is individualism itself that produces uniformity.

This is poorly understood by Americans, who tend to confuse uniformity with "conformity," in the sense of compliance with group demands. Many societies exert far more pressure on the individual to mold herself to play a sharply defined role in a total group pattern, but there is variation among these circumscribed roles. Our society gives more leeway to the individual to pursue her own ends, but since the culture defines what is worthy and desirable, everyone tends, independently but monotonously, to pursue the same things in the same way. Thus cooperation tends to produce variety, while competition generates uniformity.

The problem with individualism is not that it is immoral but that it is incorrect. The universe does not consist of a lot of unrelated particles but is an interconnected whole. Pretending that our fortunes are independent of each other may be perfectly ethical, but it's also perfectly stupid. Individualistic thinking is unflagging in the production of false dichotomies, such as "conformity vs. independence," "altruism vs. egoism," "inner-directed vs. other-directed," and so on, all of which are built upon the absurd assumption that the individual can be considered separately from the environment of which he or she is a part.

A favorite delusion of individualism—one that it attempts, through education and propaganda, to make real—is that only egoistic responses are spontaneous. But this is not so: collective responses—helping behavior, nurturance, supportiveness, the assumption of specialized roles in group tasks, rituals, or games—these are natural, not trained, even among animals. People are more *self-consciously* oriented toward others in competitive, individualistic societies—their behavior is calculated. They accommodate to others because they "want to look good, impress people, protect themselves from shame and guilt, and avoid confronting people directly. In more organic and cooperative communities people respond spontaneously to impulses that are either selfish nor unselfish, but more directly from the heart. Sometimes they look generous, sometimes grasping, but what's

important is that the behavior is to others, not an effort to produce some sort of *effect* on others. Cooperative societies are un-
assuming—it's the competitive ones that are concerned with ap-
pearances.

Individualism in the United States is exemplified by the flight
to the suburb and the do-it-yourself movement. Both attempt to
deny human interdependence and pursue unrealistic fantasies of
self-sufficiency. The first tries to overlook our dependence upon
the city for the maintenance of the level of culture we demand.
"Civilized" means, literally, "citified," and the state of the city is
an accurate index of the condition of the culture as a whole. We
behave toward our cities like an irascible farmer who never feeds
his cow and then kicks her when she fails to give enough milk.
But the flight to the suburb was in any case self-defeating, its
goals subverted by the mass quality of the exodus. The suburban
dweller sought peace, privacy, nature, community, good schools,
and a healthy child-rearing environment. Instead, he found neither the beauty and serenity of the countryside nor the stimulation of the city, nor the stability and sense of community of the
small town. A small town, after all, is a microcosm, while the
suburb is merely a layer, narrowly segregated by age and social
class. A minor irony of the suburban dream is that, for many
Americans, reaching the pinnacle of their social ambitions (own-
ing a house in the suburbs) forces them to perform all kinds of
menial tasks (carrying garbage cans, mowing lawns, shoveling
snow, and so on) that were performed for them when they occupied a less exalted status.

Some of this manual labor, however, is voluntary—an attempt
to deny the division of labor required in a complex society. Many
Americans seem quite willing to pay the price rather than engage
in encounters with workers. This do-it-yourself trend has accompanied increasing specialization in occupations. As one's job nar-
rows, perhaps, he or she seeks the challenge of new skill-
acquisition in the home. But specialization also means that one's
encounters with artisans in the home proliferate and become
more impersonal. It's no longer a matter of a few well-known
people—smiths and grocers—who perform many functions and
with whom contact may be a source of satisfaction. One finds

instead a multiplicity of narrow specialists each perhaps a stranger—the same type of repair may even be performed by a different person each time. Every relationship, such as it is, must start from scratch, and it's small wonder the householder turns away from such an unrewarding prospect in apathy and despair.

*   *   *

Americans thus find themselves in a vicious circle in which their community relationships are increasingly competitive, trivial, and irksome in part as a result of their efforts to avoid or minimize potentially irksome relationships. As the few vestiges of stable community life erode, the desire for a simple, cooperative lifestyle flows in intensity. The most seductive appeal of radical ideologies for Americans consists in the fact that all in one way or another attack the competitive foundations of our society.

Now it may be objected that American society is less competitive than it once was, and that the appeal of radical ideologies should hence be diminished. Social critics in the fifties argued that the entrepreneurial individualist of the past has been replaced by a bureaucratic Organization Man. Much of this historical drama was created by comparing yesterday's owner-president with today's assistant sales manager; certainly these nostalgia-merchants never visited a nineteenth-century company town. Another distortion is introduced by the fact that it was only the most ruthlessly competitive robber barons who survived to tell us how it was. Little is written about the neighborhood store that extended credit to the poor, or the small town industry that refused to lay off local workers in hard times. They all went under together. The meek may be blessed but they don't write memoirs.

Even if we grant that the business world was more competitive in the nineteenth century, the total environment was less so. The individual worked in a smaller firm with lower turnover in which his or her relationships were more enduring and more personal. The ideology of Adam Smith was tempered by the fact that the participants in economic struggles were neighbors and might have been childhood playmates. Even if the business world then was as "dog-eat-dog" as we imagine it, it occurred as a deviant episode in what was otherwise a more comfortable and familiar environment than the organization man can find today in or out of his office. The organization man is simply a carryover from the paternalistic environment of the family business and the company town; and the "other-directedness" of the suburban community just a desperate attempt to bring some old-fashioned small-town collectivism into the transient and impersonal lifestyle of the suburb. The social critics of the 1950s were so preoccupied with assailing these rather synthetic forms of human interdependence that they lost sight of the underlying sickness that produced them. Medical symptoms usually result from attempts made by the body to counteract disease, and attacking the symptoms often aggravates and prolongs the illness. This seems to be the case with the feeble and self-defeating efforts of twentieth-century Americans to create a viable social environment.

II. ENGAGEMENT AND DETACHMENT: GETTING INVOLVED

Many of the things we've discussed can also be traced to a compulsive American tendency to avoid confronting chronic social problems. This tendency often comes as a surprise to foreigners, who think of Americans as pragmatic and down-to-earth. But while trying to solve long-range social problems with short-run "hardware" solutions produces a lot of hardware—a down-to-earth result, surely—it can hardly be considered practical when it aggravates the problems, as it almost always does. American pragmatism is deeply irrational in this respect, and in our hearts we've always known it. One of the favorite themes of American cartoonists is the man who paints himself into a corner, saws off the limb he's sitting on, or runs out of space on the sign he's printing. The scientist of horror films, whose experiments lead to disastrously unforeseen consequences, is a more nervous version of this same awareness that the most future-oriented nation in the world shows a deep incapacity to plan ahead. We are, as a people, perturbed by our inability to anticipate the conse-

quences of our acts, but we still wait optimistically for some
magic telegram, informing us that the tangled skein of misery
and self-deception into which we have woven ourselves has vanished in the night. Each month popular magazines regale their
readers with such telegrams: announcing that our transportation
crisis will be solved by a bigger plane or a wider road, and mental
illness cured with a pill, poverty with a law, slums with a bull-
dozer, urban violence with a new weapon, racism with a goodwill
gesture. One of the most grotesque of these was an article in *Life*
some years ago showing a group of suburbanites participating in
a "Clean-Up Day," in an urban slum. Foreigners are surprised
when Americans exhibit this kind of naïveté and/or cynicism
about social problems—they don't realize that no matter what
realism we may display in technical areas, our approach to social
issues inevitably falls back on a cinematic tradition in which social problems are resolved by a gesture. Deeply embedded in
the somnolent social consciousness of the broom-wielding suburbanites is a series of climactic movie scenes in which a long
column of once surly natives, marching in solemn silence,
framed by the setting sun, turn in their weapons to the white
chief who has done them a good turn, or menace the white ad-
venturer's enemy (who turns pale at the sight), or rebuild the
missionary's church, destroyed by fire.

When a social problem persists (as they always do), those who
call attention to its continued presence are accused of "going too
far" and "causing the pendulum to swing the other way." We can
make war on poverty but shrink from the extensive changes required to stop breeding it. Once a law is passed, a commission
set up, a study made, a report written, the problem is expected
to have been "wiped out" or "mopped up." Bombs abroad are
matched by "crash programs" at home—the terminological sim-
ilarity reveals a psychological one. Our approach to transportation problems has had the effect of making it easier and easier to
travel to more and more places that have become less and less
worth driving to. Asking us to consider the manifold consequences of chopping down a forest, draining a swamp, spraying
a field with poison, making it easier to drive into an already
crowded city, or selling deadly weapons to everyone who wants

them arouses in us the same impatience as a chess problem would in a hyperactive six-year-old.

The avoiding tendency lies at the very root of American character. This nation was settled and continually repopulated by people who were not personally successful in confronting the social conditions in their mother country, but fled in the hope of a better life. By a kind of natural selection, America was disproportionately populated with a certain kind of person.

In the past we've always stressed the positive side of this selection, implying that America thereby found herself blessed with an unusual number of energetic, mobile, ambitious, daring, and optimistic persons. Now there's no reason to deny that there were differences between those who chose to come and those who chose to stay, nor that these differences must have reproduced themselves in social institutions. But very little attention has been paid to the negative side of the selection. If we gained the energetic and daring, we also gained the lion's share of the rootless, the unscrupulous, those who valued money over relationships, and those who put self-aggrandizement ahead of love and loyalty. And most of all, we gained an undue proportion of persons who, when faced with a difficult situation, tended to chuck the whole thing and flee to a new environment. Escaping, evading, and avoiding are responses which lie at the base of much that is peculiarly American—the suburb, the automobile, the self-service store, and so on.

These responses also contribute to the appalling discrepancy between our wealth and our treatment of those who cannot adequately care for themselves. In a cooperative, stable society, the aged, infirm, or psychotic person can be absorbed by the local community, which knows and understands him. He presents a familiar difficulty that can be confronted daily and directly. This situation cannot be reproduced in our society today—the same burden must be carried by a small, isolated, mobile family unit that is not really equipped for it.

But if we are forced to incarcerate those who can't function independently in our society, we ought at least to know what we're doing when we do it. The institutions we provide for those who cannot care for themselves are human garbage heaps—they both result from and reinforce our tendency to avoid confronting social and interpersonal problems. They make life "easier" for the rest of society, just like the automobile. And just as we find ourselves devising ridiculous exercises to counteract the harmful effects of our dependence upon the automobile, so the "ease" of our social technology makes us bored, flabby, and insensitive, and our lives empty and mechanical.

The Toilet Assumption

Our ideas about institutionalizing the aged, psychotic, retarded, and infirm are based on a pattern of thought that we might call the Toilet Assumption—the notion that unwanted matter, unwanted difficulties, unwanted complexities and obstacles will disappear if they're removed from our immediate field of vision. We don't connect the trash we throw from the car window with the trash in our streets, and we assume that replacing old buildings with new expensive ones will alleviate poverty in the slums. We throw the aged and psychotic into institutional holes where they cannot be seen. Our approach to social problems is to decrease their visibility: out of sight, out of mind. This is the real foundation of racial segregation, especially its most extreme case, the Indian "reservation." The result of our social efforts has been to remove the underlying problems of our society farther and farther from daily experience and daily consciousness, and hence to decrease, in the mass of the population, the knowledge, skill, and motivation necessary to deal with them.

When these discarded problems rise to the surface again—a riot, a protest, an exposé in the mass media—we react as if a sewer had backed up. We are shocked, disgusted, and angered, and immediately call for the emergency plumber (the special commission, the crash program) to ensure that the problem is once again removed from consciousness.

The Toilet Assumption isn't just a facetious metaphor. Prior to the widespread use of the flush toilet all of humanity was daily confronted with the immediate reality of human waste and its disposal. They knew where it was and how it got there. Nothing miraculously vanished. Excrement was conspicuously present in the outhouse or chamber pot, and the slops that went out the window went noticeably into the street. The most aristocratic Victorian ladies strolling in fashionable city parks thought nothing of retiring to the bushes to relieve themselves. Similarly, garbage did not disappear down a disposal unit—it remained nearby.

As with physical waste, so with social problems. The biblical adage "the poor are always with us" had a more literal meaning before World War I. The poor were visible and all around. Psychosis was not a strange phenomenon in a textbook but a familiar neighbor or village character. The aged were in every house. Everyone had seen animals slaughtered and knew what they were eating when they ate them; illness and death were a part of everyone's immediate experience.

In contemporary life the book of experience is filled with blank and mysterious pages. Occupational specialisation and plumbing have exerted a kind of censorship over our understanding of the world we live in and how it operates. And when we come into immediate contact with anything that doesn't fit the everyday pattern of our bowdlerized existence, our spontaneous reaction is to try somehow to bomb it away or flush it down the jail.

The Grappling Impulse

But to some small degree we also feel bored and uneasy with the orderly chrome and porcelain vacuum of our lives, from which so much of life has been removed. Evasion creates self-distaste as well as comfort, and radical confrontations are exciting as well as disruptive. The answering chord that they produce within us is terrifying, and although we cannot entirely contain our fascination, it's easy to project our self-disgust onto those who do the confronting.

This ambivalence is respected in the mass media. The hunger for confrontation and experience draws a lot of attention to social problems, but these are usually dealt with in such a way as to reinforce our avoidance. The TV documentary presents a tidy package with opposing views and an insinuation of progress. Printed articles and reviews give just the blend of titillation and condescension to make readers imagine that they're already "in" and need not undergo the experience itself—that they've not only participated in the novel adventure but already outgrown it. Thus the ultimate effect of the media is to reinforce the avoiding response by providing people with an effigy of confrontation. There is always the danger with these insulating mechanisms, however, that they sometimes get overloaded, like tonsils, and become carriers of the very forces they're directed against. This is a frequent event in our society today.

Closely related to this latent desire for confrontation is a wish to move in an environment consisting of something other than our own creations. Human beings evolved as organisms geared to struggle with the natural environment. Within the past few thousand years we've learned to perform this function so well that nature poses very little threat to us. Our dangers are self-made ones—subtle, insidious, and meaningless. We die from our own machines, our own poisons, our own weapons, our own despair. Furthermore, too little time has gone by for us to have adapted biologically to a completely man-made environment. We still long for and enjoy struggling against the elements, even though such activity is only occasionally meaningful.* We cross the ocean in primitive boats, climb mountains we could fly over, kill animals we don't eat. Natural disasters, such as floods, hurricanes, blizzards, and so on, provoke a cheerfulness that would seem odd if we didn't all share it. It's as if some balance between humans and nature had been restored. Like the cat that prefers to play with a ball around the obstacle of a chair leg, humans seem to derive some perverse joy from having a snowstorm force them to use the most primitive mode of transportation. It's particularly amusing to watch people following the course of an ap-

---

*The cholesterol problem gives us an illustration: one theory says that the release of cholesterol into the bloodstream was functional for hunting large animals with primitive weapons. Since the animal was rarely killed but only wounded, he had to be followed until he dropped, and this meant walking or running for several days without food or rest. The same body response would be activated today in jobs where a sustained extra effort for a week or two (to obtain a large contract, for example) is periodically required. But these peak efforts involve no physical release—the cholesterol is not utilized.

proaching hurricane affect a proper and prudent desire that it
veer off somewhere, in the face of an ill-concealed craving that it
do nothing of the kind. There is a satisfaction that comes from
relating to nature on equal terms, with respect and even deference to forms of life different from ourselves—as the Indian respects the deer he kills for food and the tree that shields him
from the sun.

Today we relate mostly to extensions of our own egos. We
stumble over the consequences of our past acts. We are drown-
ing in our own excreta (another consequence of the Toilet Assumption). We rarely come into contact with a force that is
clearly and cleanly not-us. Every struggle is a struggle with ourselves, because there is a little piece of ourselves in almost
everything we encounter—houses, clothes, cars, cities, machines, even our foods. There is an uneasy, anesthetized feeling
about this kind of life—like being trapped forever inside an air-
conditioned car with power steering and power brakes and only
a telephone to talk to. Our world is only a mirror, and our efforts
mere shadowboxing—yet shadowboxing in which we frequently
manage to hurt ourselves.

Even that part of the world that isn't man-made impinges
upon us through a symbolic network we've created. We encounter mainly our own fantasies: we have a concept and image of a
mountain, lake, or forest almost before we ever see one. Travel
posters tell us what it means to be in a strange land, the events
of life become news items before they actually happen—all experience receives preliminary structure and interpretation. Public relations, television drama, and life become indistinguish-
able.

The story of Pygmalion is thus the story of modern man, in
love with his own product. But like all discreet fairy tales, that
of Pygmalion stops with the consummation of his love. It does
not tell us of his ineffable boredom at having nothing to love but
an excrescence of himself. But we know that men who live surrounded by those whom they have molded to their desires—
from the Caliph of Baghdad to Federico Fellini—suffer from a
fearsome ennui. The minute they assume material form our fantasies cease to be interesting and become mere excreta.

Walking home one night after seeing Disney's *Fantasia* I was suddenly struck with the resemblance between the oncoming cars choking the residential street and the army of brooms that overwhelmed the sorcerer's apprentice. We aren't as wise as the sorcerer, who knew better than to waste his magic on labor-saving devices. We take after the witless and lazy apprentice in that we know how to start a magical process but not how to stop it.

III. DEPENDENCE AND INDEPENDENCE: SHARING THE WEIGHT

Independence training in American society begins almost at birth—babies are held and carried less than in most societies and spend more time in complete isolation. When a child is admonished to be a "big boy" or "big girl," this usually means doing something alone or without help (the rest of the time it means strangling feelings). Signs of independence are usually rewarded, and a child who calls too obvious attention to the fact that learning is based almost entirely on imitation is ridiculed by being called a copycat or a monkey (humans have a strange habit of projecting their uniquely human attributes onto animals).

There have been complaints in recent years that independence training is less rigorous than it once was, but once again these historical changes are hard to assess. To make one's own decisions in a simple, stable, and familiar environment requires a good deal less "independence" than in a complex, shifting, and strange one, and if the need for parental support is greater, it may cause the child to experience the independence training as more severe rather than less.

In any case, American independence training has in the past been severe relative to the rest of the world, and while this was quite consonant with the demands of adult society, the frustration of any need has its effects—one of them being to increase the society's vulnerability to social change.

An example might help clarify this issue. Ezra and Suzanne Vogel observe that Japanese parents encourage dependency as

actively as American parents push independence, and that
healthy children and adults in Japan rely heavily on others for
emotional support and decisions about their lives. A degree of
dependence on the mother which in America would be considered "abnormal" prepares the Japanese for a society in which far
more dependency is expected and accepted than in ours. The
Japanese firm is highly paternalistic and takes a great deal of responsibility for making the employee secure and comfortable.
The Vogels observe, however, that just as the American mother
has always complained about the success of her efforts and felt
that her children were *too* independent, so the Japanese mother
tends to feel that her children are too *dependent,* despite the fact
that she has trained them this way.[2]

What I'm trying to say is that regardless of how congruent
child training and adult norms may be, any extreme training pattern will produce stresses for the individuals involved. And just
as the mothers experience discomfort with the effects of these
patterns, so do the children, although barred by cultural values
from recognizing and naming the nature of their distress, which
in our society comes from a desire to relinquish responsibility for
decision-making in daily life. Since democratic values are deeply
held, the temptation to abdicate self-direction usually expresses
itself in subtle, nonpolitical ways. *Choice* is the major issue:
Americans make more conscious choices per day, with fewer
givens, more ambiguous criteria, and less environmental or social stability, than any people in history. One of the most important (and certainly the most enduring) aspects of the countercultural revolt of the sixties was a defiant refusal by young people
to assume "adult" status—to take control of their own lives, make
plans and decisions, support themselves, and so on. Although it
necessarily took a rather passive form, it was as much an attack
on the status quo as any campus protest or street demonstration.

The Oral Culture

The burden of decision-making is probably at the root of the extreme orality in American culture. Much has been made of our
having become, since World War II, a "consumer society," and
this should be taken in a very literal sense. The advertising industry caters heavily to our groaning need to be fed, while at the

same time it aggravates it further by burdening us with more and more esoteric choices. Every possession we acquire loads us with more responsibilities and cares, to which we respond by retreating into infant orality and "consuming," thereby acquiring still more possessions. This vicious circle has inspired our economy for thirty years now, so that we've come to view an "expanding" economy as an inevitability, like the expanding waistline of middle age. In some sane corner of our minds we've always known that something was wrong with this thinking—that chronic inflation was as bad for our economy as for our stomachs—but until recently we've tried to keep that thought from surfacing.

The American love of bigness is in itself a sign of orality—we can never have enough, always want more, always fear that they'll come and take it away before we've finished gorging ourselves. Even our sexuality is primarily oral, with sexual attractiveness often defined, like everything else, in terms of size: mammoth mammaries for women, prodigious penises for men— a preoccupation that has little to do with pleasurable sensation and much to do with infantile feelings of deprivation.

The columnist David Wilson once suggested[3] that "I Can't Believe I Ate the Whole Thing" (part of a TV commercial for Alka Seltzer, which Wilson calls our national beverage) would be a "perfect slogan for a somewhat shamefaced nation which, with about one-sixteenth of the earth's population, manages to consume about a third of its energy and half its resources" (and which spends 2 billion dollars annually on dieting).

This orality is stimulated not only by overchoice but also by the requirements of individual control. In stable societies the control of human impulses is usually a collective responsibility. The individual is not expected to be able at all times to keep his or her impulses from breaking out in ways disapproved by the community. But this matters very little, since the group is always near at hand to stop or shame or punish should one forget oneself.

Control from Within

In more fluid, changing societies controls are more apt to be internalized—to depend less on external enforcers. This has long been true of American society—Tocqueville observed in 1830, for example, that American women were much more independent than European women, freer from chaperonage, able to appear in what a European would consider "compromising" situations without any sign of sexual involvement.

Chaperonage is the simplest way to illustrate the difference between external and internalized controls. In chaperon cultures—such as traditional Middle-Eastern and Latin societies— it simply did not occur to anyone that a man and woman could be alone together and not have sexual intercourse. In modern America, by contrast, there is no situation in which a middle-class man and woman could find themselves that would make sexual intercourse automatic. Hollywood comedies have been exploiting this phenomenon—well past the point of exhaustion and nausea—for the past forty years. Americans are virtuosi at controlling sexual expression, and the current relaxation of sexual norms in no way changes this. It causes difficulties, however, when the two systems come into contact: a woman in a bikini means one thing in America, another in Baghdad. Even Americans tend to consider some situations inherently sexual: if a woman from another culture came to an American male's house, stripped, and climbed into bed with him, he would assume she was making a sexual overture and would be rather indignant to find she was merely expressing polite friendship according to her native customs.

But we don't need to go outside our own culture to find such contrasts. It isn't uncommon for college students who know each other only casually to share a bed without making love, yet without ruling out the possibility that they might. This is very confusing to their parents' generation, for whom such an episode would be automatically sexualized, and thus would be seen as an example either of Herculean chastity or sexual neuroticism. Some parents, in fact, upon hearing of such experiences, use them to deny their child's sexuality altogether—blithely assuming that because one night's encounter was unconsummated, all are.

But how are internalized controls created? We know that they're closely tied to techniques of discipline that emphasize

rewards and reasoning rather than physical punishment and deprivation, so that instead of merely learning to avoid punishment,
the child comes to incorporate parental values as his or her own
in order to avoid losing parental love and approval. When more
intimidating techniques prevail, the child becomes like the inhabitants of an occupied country who obey to avoid getting hurt,
but disobey whenever they think they can get away with it: the
child doesn't have a strong emotional commitment to the parents, or fear losing their love.

For internalized controls to develop in the child, love and discipline must emanate from the same source. When this happens
it isn't merely a question of avoiding the punisher: the child
wants to anticipate the displeasure of the loving parent, wants to
be like the parent, and hence absorbs the values and attitudes of
the parent. He or she wants to please, not placate, and because
she has taken the parent's attitudes as her own, pleasing the parent comes to mean making her feel good about herself.

Under stable conditions external controls work perfectly well.
Everyone knows her own place and her neighbor's, and social
deviations are quickly countered from all sides. When conditions
fluctuate, norms change, people move frequently and are often
among strangers, this will no longer do. One cannot take her
whole community with her wherever she goes, and the rules
differ from place to place. A mobile individual must travel light,
and internalized controls are portable and transistorized, as it
were. In a stable community, for example, two youths who get
into a fight will be held back by their friends—they depend on
this restraint and can abandon themselves to their passion,
knowing that it won't have harmful consequences. But where
people move among strangers it becomes necessary to have
other mechanisms for handling aggression. In situations of high
mobility and flux a man must have a built-in readiness to feel
himself responsible when things go wrong.

Most modern societies are a confused mixture of both systems, a fact that enables conservative spokesmen to attribute rising crime rates to permissive child-rearing techniques, while liberals point to poverty and authoritarian schooling. Actually, it's a
little misleading to call the guilt-inducing child-rearing methods

of middle-class parents "permissive." Misbehavior in a working class child might meet with a cuff, possibly accompanied by some non-informative response like "stop that!" But it may not be clear to the child which of the many motions he is now performing "that" is; and, indeed, "that" may be punished only when the parent is feeling irritable. A child would have to be very intelligent to form a moral concept out of a hundred irritable stop-thats. What he usually forms is merely a crude sense of when the parent is to be avoided. The self-conscious, verbal, middle-class parent, on the other hand, usually feels that discipline should relate to the child's act, not to the parent's emotional state, and is careful to emphasize verbally the principle involved in the misbehavior ("it's bad to hit people" or "we have to share with guests"). Concept-formation is made easy for the middle-class child, and she tends to think of moral questions in terms of principles.

As the child grows older this tendency is reinforced by her encounters with new groups that have different norms. In a mobile society, no rule has absolute validity because people are aware of competing moral codes. The mobile middle-class child therefore tends to evolve a system of meta-rules—that is, rules for assessing the merits of these competing codes. The meta-rules tend to be based on the earliest and most general principles expressed by parents, such as prohibitions on violence against others, equality, fairness, and so on. This ability to treat rules in a secular fashion while maintaining a strong moral position is baffling to those whose control mechanisms are more primitive, but it presupposes a powerful and articulate conscience. Such a person is capable of exposing herself to violence-arousing situations without losing control and while maintaining a moral position, which seems inconceivable to an uneducated policeman whose own impulses are barely held in line by a jerry-built structure of poorly articulated and mutually contradictory moral absolutes. Hence policemen often interpret radical middle-class activism as a hypocritical mask for simple delinquency.

The point of this long digression, however, is that internalization is a mixed blessing. It may enable one to get his head smashed in a good cause, but the capacity to give oneself up completely to an emotion is almost lost in the process. Where internalization is high there is often a feeling that the controls themselves are out of control—that emotion cannot be expressed when the person would like to express it. Life is muted, experience filtered, emotion anesthetized or its discharge incomplete. Modern efforts to shake free from this system of hypertrophied control include the humanistic psychology movement, the use of psychedelic drugs, and attempts to re-establish external systems of direction and control, such as astrology and various Eastern religious disciplines. The simplest technique, of course, would be to establish a more authoritarian social structure, which would relieve the individual of the great burden of examining and controlling his own responses. He could become as a child: lighthearted, spontaneous, and passionate, secure in the knowledge that others would prevent his impulses from causing harm.

This solution is blocked by democratic values and the social conditions that foster them (complexity, fluidity, change). But the desire plays a significant part in conventional reactions to radical minorities, who are all felt to be seeking the abandonment of self-restraints of one kind or another while at the same time demanding *more* responsible behavior from the establishment. This is both infuriating and contagious to middle-class conservatives, who would like very much to do the same. Their call for "law and order" (that is, more *external* control) is an expression of that desire as well as an attempt to smother it. This conflict over dependency and internalization also helps explain why official American anti-communism always lays so much stress on the authoritarian (rather than the socialistic) aspects of communist life.

Individualism

These three needs—community, engagement, dependency— are suppressed in our society out of a commitment to individualism. The belief that everyone should pursue her own destiny autonomously has forced us to maintain an emotional detachment from our social and physical environment and aroused a vague guilt about our competitiveness and indifference to others. For our earliest training in childhood does not stress competitiveness, but cooperation, sharing, and thoughtfulness—it's only later that we learn to reverse these priorities. Radical challenges to our society always tap a confused responsive chord within us that's far more disturbing than anything going on outside. They threaten to reconnect us with each other, with nature, and with ourselves, a possibility that's thrilling but terrifying—as if we had grown a shell-like epidermis and someone was threatening to rip it off.

Individualism is rooted in the attempt to deny the reality of human interdependence. One of the major goals of technology in America is to "free" us from the necessity of relating to, submitting to, depending upon, or controlling other people.* Unfortunately, the more we have succeeded in doing this, the more we have felt disconnected, bored, lonely, unprotected, unnecessary, and unsafe.

Individualism has many guises: free enterprise, self-service, academic freedom, suburbia, permissive gunlaws, civil liberties, do-it-yourself, oil-depletion allowances. Everyone values some of these guises and condemns others, but the principle is widely shared. Criticisms of our society since World War II have almost all embraced individualism and expressed fears for its demise. Most of these critics have failed to see the role of the value they embrace so fervently in generating the phenomena they so detest. For human beings *are* interdependent—they can only *pretend* not to be. As a way of looking at the world, individualism is extremely cumbersome—when things go wrong we always have to waste a lot of valuable time trying to decide whose fault it is, since we start with the silly assumption that everyone and everything is separate and autonomous.

---

*The peculiar germ-phobia that pervades American life (and supports several industries) also owes much to this. We have carried the fantasy of individual autonomy so far that we imagine each person to have his own unique species of germs, which must therefore not be mixed and confused with someone else's. We're even disturbed at the presence of the germs themselves: although many millions inhabit every healthy human body from the cradle to the grave we regard them as trespassers. We feel that nature has no business claiming a connection with us, and perhaps one day we will prove ourselves correct.

The most sophisticated apologist for individualism is David Riesman, who at least recognizes that uniformity and community are not the same thing. Perhaps the definitive and revealing statement of what individualism is all about is his: "I am insisting that no ideology, however noble, can justify the sacrifice of an individual to the needs of the group."[4]

Whenever I hear such sentiments I recall Jay Haley's discussion of the way the families of schizophrenics communicate. He points out that people who communicate necessarily govern each other's behavior—set rules for each other. But an individual may try to avoid this human fate—to become independent, uninvolved: ". . . he may choose the schizophrenic way and indicate that nothing he does is done in relationship to other people." The family of the schizophrenic establishes a system of rules like all families, but also has "a prohibition on any acknowledgment that a family member is setting rules. Each refuses to concede that he is circumscribing the behavior of others, and each refuses to concede that any other family member is governing him." The attempt, of course, fails. "The more a person tries to avoid being governed or governing others, the more helpless he becomes and so governs others by forcing them to take care of him."[5] In our society as a whole this caretaking role is assigned to technology, like so much else.

Riesman overlooks the fact that the individual is sacrificed either way. If he is never sacrificed to the group, the group will collapse and the individual with it. Part of the individual is, after all, committed to the group. Part of him wants what "the group" wants, part does not. No matter what is done some aspect of the individual will be sacrificed.

An individual, like a group, is a motley collection of ambivalent feelings, contradictory needs and values, and antithetical ideas. He is not, and cannot be, monolithic, and the modern effort to pretend otherwise is not only delusional and ridiculous, but also acutely destructive, both to the individual and to society.

The reason a group needs the kind of creative deviant Riesman values is the same reason it needs to sacrifice her: the failure of the group members to recognize the diversity and ambivalence within themselves. Since they have rejected parts of themselves, they not only can't tap those resources but also can't tolerate their naked exposure by others. The deviant is an attempt to remedy this condition. She comes along and tries to provide what is "lacking" in the group (that is, what is present but suppressed). Her role is like that of the mutant—most are sacrificed but a few survive to save the group from itself in times of change. Individualism is a kind of desperate plea to save all mutants, on the grounds that we don't know what we are or what we need. As such it's horribly expensive—a little like setting a million chimps to banging on a typewriter on the grounds that eventually one will produce a masterpiece.

But if we abandon the monolithic pretense and recognize that any group sentiment expresses some part of our being (but only a part), then the deviant prophet is unnecessary since she exists in all of us. And should she appear, it will be unnecessary to sacrifice her since we've already admitted that what she's saying is true. And in the meantime we would be able to exercise our humanity, governing each other and being governed, instead of encasing ourselves in the leaden armor of our technological schizophrenia.

TWO

Kill Anything That Moves

*I am afeard there are few die well that die in a
battle; for how can they charitably dispose of any
thing, when blood is their argument?*

SHAKESPEARE

*Am I a spy in the land of the living,
That I should deliver men to Death?*

MILLAY

*The whole wide human race is taking far too
much methedrine.*

LEITCH

The war in Vietnam is over, along with the great national debate
about its morality. The only question remaining is whether we
learned anything from it. The United States has been involved
in dubious military escapades before but never on such a huge
scale. It seems certain that future historians will date our decline
as a world power from the moment when we began to squander
our energy and resources in this futile orgy of destruction.

It was a strange piece of stubbornness. We watched carefully
while France failed, and then repeated all her mistakes on a
larger scale each failure leading to further escalation. When the
war in the South floundered, military lobbyists demanded per-
mission to bomb the North. When the bombing proved ineffectual, they insisted that it be expanded, and so on. It's a standard
Pentagon tactic for financing weapon boondoggles as well as war:
"I've lost your $500 gambling; give me a thousand to win it back
or you'll never see it again." So far it's worked every time, to the
continued impoverishment of American life.

The escalation of failure has respectable but inauspicious precedents. When Athens was unable to defeat Sparta she invaded Syracuse, and extinguished herself as a dominant political power. The United States has similarly exhausted itself in Vietnam. Economists are reluctant to make the connection between the war and our economic miseries, but wars sap energy, and the distribution of energy is what economics is all about. People say war "stimulates the economy," but this is true only when everyone is involved in it. Sucking energy from domestic programs to pour into foreign destruction merely concentrates wealth in the hands of those few venal enough to profit from the anguish of others. Economic statistics and doubletalk cannot hide the fact that for ten years we squandered our wealth, power, energy, and human resources in an act of meaningless hatred.

How did it come about? The official middle-of-the-road position on Vietnam is that our involvement was merely a kind of bumbling stupidity, a well-intentioned naïveté: "I'm dreadfully sorry, but I seem to have burned down your house and slaughtered your children in my clumsiness. I thought I was doing it for your own good, but I seem to have blundered." Long and careful books have been written to show how, step by step, innocent and goodhearted men came to immerse themselves in human blood. The United States as international schlemiel.

I'm afraid my own gullibility isn't up to this yarn. Not that I believe the designers of the Vietnam war were self-consciously malevolent. I'm sure they managed to persuade themselves, like other mass killers of history, that they were doing the right thing. Some of them were probably considered kindly men by their families and intimate friends. Men with the lust for power have a rather eerie ability to compartmentalize their actions.

I'm not really interested in the individuals involved. Vietnam was a national endeavor, and insofar as we all failed to stop it, we all colluded in it. What concerns me is what motivated the nation as a whole to engage in this demented adventure. Stupidity explains nothing. We're not so dumb that we can't see a dollar bill when it's waved in front of us. What led us to con ourselves into thinking there could be anything moral, practical, or heroic about trying to slaughter an entire nation of underweaponed peasants, a nation that humbled two nuclear powers without having even a viable air force? The *motivation* for this is what concerns me. What was there in the Vietnam war that was satisfying enough to blind us to its folly? To solve this we need to talk a little about violence.

When Do We Approve of Violence?

The sixties brought an end to our illusion that we are a peaceful people. Americans have always admitted being lawless compared with Europeans, but this was attributed to our recent frontier history. High crime rates were merely a sign of youthful high spirits ("America is all boy!"), and a subtle contempt tainted our respect for the law-abiding English. Today the chuckle is gone, for our violence has become less casual. As a nation we seem obsessed with it, to the point where the bulk of our population spends hours each evening looking at it.

Violence has always been acceptable to middle-class Americans when it was (1) nonpolitical and (2) confined to poor neighborhoods. These two conditions are related in that they both protect the advantaged from the anger of the disadvantaged. Nonpolitical violence is unfocused, confused. When it begins to take on political coloring, it's like the difference between throwing an ineffectual tantrum and punching someone in the nose. We've always had race riots, campus riots, ethnic clashes, gang wars, and shootouts, but these have for the most part lacked any political thrust. Even the urban riots of the sixties were pretty much undirected explosions, but the mere knowledge that blacks *felt* angry about white exploitation, even though they were burning their own homes, led to a new concern about urban violence among middle-class whites who had been unruffled by the race riots of the past. Or, to take a more trivial example, campus riots and fraternity pranks in the fifties produced property damage and injuries as severe as any campus demonstration of the sixties. Harvard University, for example, had annual spring riots that occasionally caused bloody encounters with local police. Many of the nation's most conservative leaders have been the heroes of tales of epic vandalism, but jail sentences were out of the question since they might "spoil a young man's career." Such incidents

were considered venial—mere boisterousness on the part of
young men who fundamentally accepted the system and who
only harmed working-class people or caused damage that
wealthy parents could pay for. In the protests of the sixties authority was confronted and questioned, and even when no one
was hurt and no damage done, those in advantaged positions who
had their routines disrupted and assumptions questioned felt as
if a psychological violence had been done to them.

For the same reason violence never perturbed middle-class
Americans when it was confined to the world of the oppressed,
to be drawn upon only when needed. The poor have seldom enjoyed the benefits of "law and order," but as long as the mayhem
never spilled over into "good neighborhoods" it aroused little
concern. So long as our society was decentralized, the chronic
violence in slums and depressed rural areas didn't disturb the
society as a whole. Barroom brawls, gang wars, and lynchings
were faraway events. But the mass media have flooded out local
boundaries and forced the middle class to become aware of what
it's like to live in fear. It isn't so much the *increase* in violence
that upsets middle-class Americans as the *democratization* of violence: the poor and black have become less willing to serve as
specialized victims of pseudo-legal violence from middle-class
whites and illegal violence from each other. The demand for law
and order is a demand to return to the days when the advantaged
groups in our society held a monopoly of this scarce commodity.

The same point can be made about crimes against property,
since our legal system has a class bias. The ways in which the
rich steal from the poor are rarely defined as crimes and even
when they are, enforcement is lenient; our great fortunes were
built on such theft. Most judges and law enforcement officials
seem to feel that it isn't right for middle-class people to go to jail
just for stealing. While a poor man may get several years for a
hundred-dollar holdup, a rich man will seldom get more than a
few months for stealing a hundred thousand. Judges seem to feel
that public officials and corporation executives are thieves and
swindlers by profession and hence entitled to leniency when
caught, like children who steal candy. They argue that the rich
and powerful are punished enough by losing their jobs and being
publicly humiliated, whereas the poor are so humiliated already

they have nothing to lose. A month or two in jail for a rich man, even if he has stolen millions, betrayed the public trust, and caused the deaths of innocent consumers, is viewed as cruel and unusual punishment. Thus the Watergate conspirators are by and large free to embark on lucrative second careers as raconteurs, bible-thumpers, memoir-writers, and experts on public morality. The biggest conspirator of all was punished by being given hundreds of thousands of poor people's dollars, along with servants, guards, and other special privileges. And people still felt sorry for him.

So if property crime rates are rising, this may only reflect an increase in the democratization of larceny, a trend attributable in part to the success of advertisers in convincing the poor that only the possession of material goods can satisfy their various social, sexual, and moral requirements.

It's almost impossible to figure out whether in fact crimes of any sort are on the increase. Fluctuations have more to do with fashions in the reporting of crime than with crime itself, and statistics are often manipulated to create certain effects. Spurious increases are good for getting more appropriations, and politicians have often thundered forth about crime in the streets when it was in fact in sharp decline.

But whatever the truth about the amount of domestic crime and violence, it has certainly become more visible and less anonymous, and our preoccupation with it has certainly grown. Whatever the age group, social class, or educational level, Americans seem to be most entertained by watching people get killed, bludgeoned, or mutilated. If anything, the intellectuals seem to like their brutality rawer than the plainer folks—witness the current vogue of Peckinpah. Films that used to be considered nightmare bait for children are now virtually all they can see, and when they come to make their own dramatic productions, unpunished violence is the rule. Whether it's the good guys or the bad guys, whoever is most violent and bloody wins the day.

Violence at a Distance

But all domestic violence, whether reality or fantasy, pales before the violence we have created outside our borders. Indeed, it seems reasonable to suggest that this violence plays some role

in all other forms, for don't the leaders of a nation set the tone
for others? And if the leaders cannot abstain from the perpetra-
tion of violence and brutality, who will be able to?

Curiously enough, our willingness to acknowledge domestic
violence has never extended to the international sphere. Included in our country-bumpkin self-image is the notion that we
always became embroiled in foreign conflicts against our will,
seduced from our peaceful pursuits. Expansionist drives against
Mexico and Spain were glossed over, along with our uniquely
bloody civil war, our brutal suppression of Philippine independence, and our strong-arm tactics in Latin America. The Vietnam
war—despite a whole new vocabulary of self-deception ("advis-
ers," "escalation," "pacification")—unraveled this illusion.

What first began to disturb thoughtful Americans about Vietnam was the prevalence of genocidal thought patterns in our approach to the conflict. I don't just mean remarks by wild-eyed
generals about "dropping a nuke," or frustrated soldiers talking
about "paving the country over from one end to the other." Official policy was expressed in more restrained language, but the
euphemisms could not entirely hide the same genocidal assumptions. "Rooting out the infrastructure," for example, meant that
you no longer killed only soldiers carrying weapons but every
civilian who might be related to or sympathetic to those soldiers.
Since there is no way of telling this at a glance in a civil war, it
simply meant killing every civilian around.

In previous wars casualties were a by-product of the goals for
which the war was being fought. Even the Nazis were primarily
interested in acquiring territory and making converts. But for us
the body count became an end in itself: each day we tallied up
how many killings we had achieved (ignoring, in the process,
how many enemies we had made among the neutral living). The
implicit assumption of the enemy-fatality statistics was "so many
today, so many tomorrow, and one day we will have killed all the
Communists in the world and live happily ever after." Defoliation, napalm, and cluster bombs were designed to exterminate a
population, not to win ground, liberate, convert, or pacify.

Media reports reflected this intent. A soldier boasted in the
press of killing over two hundred people. Another, discovered to

be underaged, protested in a newspaper headline: I CAN KILL AS WELL AS ANYBODY. And when the bombing of North Vietnam was resumed after a lull in January, 1966, a Boston newspaper carried the cheery headline, BOMBS AWAY!

Americans have always been genocidal: witness our systematic extermination of the Indian, the casual killing of blacks during and after slavery, and our willingness to drop the atomic bomb on a large civilian populace (we are, after all, the only people ever to have used such a weapon). We have long had a disturbing tendency to see non-whites—particularly Orientals—as nonhuman, and to act accordingly. While communist countries usually enjoy the benefit of our fantasy that their people are ordinary humans enslaved by evil despots and awaiting liberation. When some event—such as the Bay of Pigs fiasco—disconfirms this fantasy, we are simply bewildered and turn our attention elsewhere. But the same disconfirmation in a nonwhite country like Vietnam just made their people candidates for extermination.

Every society that has achieved pre-eminence in the world has shown a remarkable capacity for brutality and violence—you don't get to be the bully of the block without using your fists. Vietnam was unusual only because it occurred in the face of some inhibiting factors—the practical and moral lessons of several decades. We knew from World War II, for example, that military force is ineffectual in changing attitudes, that air power is cruel but futile against civilian populations, that colonial expeditionary forces are impotent against organized indigenous movements of any size, and that military dictators will not and cannot broaden their base of support. We had seen France fail, both in Vietnam and in Algeria. We had helped establish international principles in the U.N., at Geneva, and at Nuremberg, which we then violated. We lived in a society in which the cruelties of war were exposed in every living room through mass media. We maintained that every human life was a thing of value. Yet we engaged in the mass slaughter of innocent persons by the most barbarous means possible and showed few qualms about it. Even the resistance to the war was largely concerned with the expense of conducting it, and the loss of *American* lives. Since we are no longer crude frontiersmen or hillbillies, what led us to condone such savagery?

Before describing the varieties of extermination we practiced in Vietnam, let me dispose of one objection. For some people, the fact that a man has been defined as an enemy justifies whatever horror we want to inflict on him, and nothing in what follows will be viewed by these readers as worthy of note. Unfortunately, in Vietnam it was hard to determine who the enemy was. We were repeatedly trapped in our own rhetoric on this matter—first by portraying ourselves as aiding a friendly Vietnamese majority against an alien and sinister minority. This created the expectation that villages "liberated" from the Viet Cong would welcome us with open arms, as Paris did in World War II. When it turned out that the villagers weren't pleased to be rescued from their husbands, brothers, sons, and fathers, we burned their villages and destroyed their crops, and began to give increased emphasis to the idea of outside agencies, particularly the North Vietnamese. We attacked North Vietnam in part because we were unwilling to admit we were fighting the people of South Vietnam.

Fortunately, the Air Force didn't deceive itself when it came to the welfare of its downed pilots, who were advised that when hit, they should try to crash into the sea, since "everybody on the ground in South or North Vietnam (when you [float] down in a parachute, at least) must be considered an enemy." Pilots were also briefed never to bad-mouth Ho Chi Minh, even in Saigon, since he was a national hero. Yet knowing this, and knowing that Arvin troops regularly smuggled or abandoned ammunition to the Viet Cong, we seldom drew the obvious conclusion—although our troops admired the Viet Cong, wondered why they kept fighting against such overwhelming technological superiority, and wished they were allies: "If we had them on our side, we'd wrap up this war in about a month."[1]

The examples that follow, then—drawn mostly from Frank Harvey's account—are not concerned with an armed enemy force but with an entire populace. If a farmer in the Mekong Delta shot back when he was bombed and strafed, he was retroactively defined as "Viet Cong."

The Joy of the Hunt

American pilots learned their trade in the Delta, where there were no trees for the peasants to hide under, and no anti-aircraft fire. It was so safe for Americans that one pilot described it as "a rabbit shoot." The young pilot learned "how it feels to drop bombs on human beings and watch huts go up in a boil of orange flame when his aluminum napalm tanks tumble into them. He gets hardened to pressing the firing button and cutting people down like little cloth dummies, as they sprint frantically under him." If he was shot down, there were so many planes in the area that his average time on the ground (or in the sea) was only eleven minutes. Thus it was a very one-sided war—as Frank Harvey (the air force's historian of the war) remarked, the Vietnamese had about as much chance against American air power as we would have against spaceships with death rays.[2]

This training prepared American pilots for the genocidal pattern of the overall war. It did not prepare them, however, for the slightly more equal contest of bombing North Vietnam in the face of anti-aircraft fire, where planes were lost in huge numbers and downed pilots were captured by the enemy. American pilots were most anxious to bomb North Vietnam until they had actually experienced the ground fire, at which point their enthusiasm waned dramatically. It became difficult, in fact, to man these missions. According to Harvey, the Tactical Air Command in Vietnam lost a squadron of pilots a month for *noncombat* reasons. Killing in a dubious war is apparently much more palatable than getting killed, and Americans are not used to fighting with anything like equal odds (imagine our outrage if the North Vietnamese had bombed *us*). In the Delta, pilots seemed surprised and almost indignant when their massive weaponry was countered with an occasional rifle shot. As Robert Crichton points out, American pilots in Vietnam became so accustomed to 1000 to 1 firepower odds that they began to feel it was their inherent right to kill people without retaliation.[3]

A favorite theme in animated cartoons is the man who is bothered by a small but persistent insect—a mosquito, termite, or fly. He starts by trying to swat it or step on it but fails. He then

escalates to insecticide sprays, guns, ten-ton weights, and so on.
But after each carefully planned attempt at extermination the
persistent little buzzing or gnawing sound is still heard, and the
man is driven to greater and greater paroxysms of rage. Finally,
in desperation, he either sets fire to his house or blows it up with
explosives—still to no avail. The message seems to he that you
can't stamp out life, no matter how much power you have.
Whenever I saw Johnson or Nixon talk about Vietnam I always
thought of that frenzied man and the indomitable insect.

The administration of extermination in the Delta was highly
decentralized. Decisions were made by forward air controllers
(FACs) who flew about looking for signs of "guerilla activity"
(which in most cases meant "life"). As Harvey reported, "They
cruise around over the Delta like a vigilante posse, holding the
power of life and death over the Vietnamese villagers living be-
neath." The weapons they could call upon had an unfortunate
tendency to kill indiscriminately. There was napalm, which
rolled and splattered about over a wide area, burning everything
burnable that it touched, suffocating those who tried to escape
by hiding in tunnels, pouring in and incinerating those who hid
in family bomb shelters under their huts. Napalm was a favorite
weapon, according to Harvey, and was routinely used on rows of
houses, individual farms, and rice paddies. "Daisy cutters," or
bombs which exploded in the water, were also used against peasants hiding in rice paddies. White phosphorus bombs were another incendiary used, and Harvey saw a man in a civilian hospital with a piece of phosphorus in his flesh, still burning.
Harvey considered the deadliest weapon of all to be the Cluster
Bomb Units (CBUs), which contained tiny bomblets expelled
over a wide area. With this device a pilot could "lawnmower for
considerable distances, killing or maiming anybody on a path
several hundred feet wide and many yards long." The CBUs
were particularly indiscriminate since many had delayed action
fuses which went off when the "suspect," whose appearance pro-
voked the FAC observer to trigger off this holocaust, was far
away, and the victims being "lawnmowered" were children playing about in what they thought was a safe area, or peasants going
about their daily work. Victims who survived sometimes under-

went unusual surgery—if hit in the abdomen, it had to be slit from top to bottom and the intestines spilled out onto a table and fingered for fragments. With one type of CBU a plane could shred an area a mile long and a quarter of a mile wide with more than a million steel fragments. It was difficult for many to reconcile this kind of indiscriminate killing with speeches about "winning the hearts and minds of the Vietnamese people."[4]

The initiative granted the FACs amounted to a mandate for genocide. If a FAC saw nothing suspicious below, he was entitled to employ "Recon by smoke" or "Recon by fire." In the first case, he dropped a smoke grenade, and if anyone ran from the explosion, they were presumed guilty and napalmed (if they ran into their house) or machine-gunned (if they took to the rice paddies). "Recon by fire" was based on the same principle except that CBUs were used instead of smoke grenades, so that if the victims did *not* run they were killed anyway. These techniques were like the ducking stool used centuries ago to test witches: if the woman was not a witch, she drowned—if she did not drown, this proved she was a witch and she was burned to death.

Bureaucratic Violence

It is the efficiency of the slaughter that impresses us, and the at times bewildering overkill—dropping bombs on individuals or using multi-million-dollar planes to "barbecue" peasant huts. When a lone farmer standing in a field managed to hit one of these overarmed pilots with a rifle shot, no reasonable bystander would have been able to stifle a cheer. But the more usual result was for the upstart to be shredded by machine-gun bullets (fired at the rate of 100 rounds per *second*) and literally to disintegrate into a pile of bloody rags. A "Huey" pilot described one such killing: "I ran that little mother all over the place hosing him with guns but somehow or other we just didn't hit him. Finally he turned on us and stood there facing us with his rifle. We really busted his ass then. Blew him up like a toy balloon." (The Huey gunship is a three-man helicopter equipped with six machine guns, rockets, and grenade-launchers).[5]

Harvey met a few FACs, however, who didn't enjoy killing civilians. One advertised his feelings and was relieved of duty.

Another, who had learned the lesson of Nuremberg, questioned an order to shell a peaceful village filled with women and children. When the order was reaffirmed, he directed the artillery fire into an empty rice paddy. Bomber pilots, however, were protected from such awareness by remoteness from their targets. B52 bombers, flying from Guam over 2500 miles away, or from Thailand, dropping bombs from 40,000 feet so that they could not be seen or heard from below, could wipe out an entire valley. In one of these "saturation" or "carpet" raids, fifty square miles of jungle would suddenly explode into flame without warning from a rain of fire bombs. These raids were frequent, and in the areas hit, nothing lived, animal or human, friend or enemy. It was almost as effective on plants and animals as defoliation, which killed three hundred acres in four minutes (the motto of the defoliators, or "Ranch Hands" was "Only You Can Prevent Forests"), probably not much more expensive (it cost almost $2 million to keep a single plane defoliating for twenty-four hours), and a great deal more inclusive.[6]

When American atrocities are discussed, the answer is often given that the Viet Cong also committed atrocities. This is a little like saying that when an elephant steps on a mouse, the mouse is an aggressor when it bites the elephant's foot. A terrorist bomb is not equivalent to a B-52 raid, nor is the sadistic murder of a captured FAC (naturally the most hated of fliers) the equivalent of a CBU drop. With our overwhelming arsenal of grotesque weapons should have gone some minimal trace of responsibility. The Viet Cong were fighting for their existence, while our pilots in the Delta were amusing themselves with impunity—their merry euphemisms like "hosing" and "barbecuing" expressed this freedom.

Implicit in the equal atrocity argument was the assumption that American lives are precious and other peoples' lives are of no more account than ants. The fantastic disproportion in firepower (". . . it is a little exaggerated . . . We're applying an $18,000,000 solution to a $2 problem. But, still, one of the little mothers *was* firing at us.") is justified in terms of saving American lives. At times Harvey seems to be describing some kind of aristocratic adventure: when an air force major was shot down over

North Vietnam, so many American planes were sent out to rescue him they had difficulty avoiding collisions. But not all of the excess could be attributed to protective concern: when a Huey gunship emptied its ammunition into total darkness ("nobody will ever know if we hit anything but we certainly did a lot of shooting"), or a B-52 rained bombs all over a forest in the hope that perhaps some Viet Cong were hiding in it, this could hardly be defined as saving American lives. It was simply gratuitous aggression, taking a form that owed much to the Toilet Assumption.

The Privileged Machine

Furthermore, the excessive American firepower and its more grisly manifestations often backfired, and destroyed those same expensive lives they were supposed to protect. Captured CBUs were made into booby traps and blew off American limbs. A large supply of our "Bouncing Betty" mines (so called because they were made to leap up and explode in the face) abandoned to the Viet Cong by the Arvins caused "sickening" casualties to our own troops. Our planes collided because there were so many. Dragon ships were melted by their own flares. Fliers were endangered when the Navy and the Air Force tried to "out-sortie" each other. We sometimes napalmed our own troops. After the *Forrestal* disaster a flier expressed momentary repugnance at having to drop napalm on North Vietnam after seeing what it had done to our own men. [7]

When all's said and done, American lives, while accorded an extraordinary value compared to Vietnamese civilians, still took a back seat to the death-dealing machinery they served. Aircraft carriers, for example, are careless of human life even under the best of conditions, remote from the field of battle. Planes often disappeared in the sea with their pilots (the cost of planes lost through landing and take-off accidents would have financed much of the poverty program), men were ignited by jet fuel, or devoured by jet engines, or run over by flight deck equipment, or blown into the sea by jet winds, or cut in half by arresting cables, or decapitated by helicopter blades. Even safety devices seemed to be geared less to human needs than to the demands of the machinery: pilots ejected from F-4s regularly suffered broken backs or other severe spinal injuries.[8] The arguments about Viet Cong atrocities and saving American lives became ludicrous in the face of the daily reality of America's life-destroying technology.

What enables civilized humans to become brutalized in this way? Why weren't more of them sickened and disillusioned, as men have often been in the past when forced to engage in one-sided slaughter?

There were really two different types of human extermination in Vietnam, and they require different explanations. First, there was extermination at close range, in which the killer could see and enjoy the blood he shed. Second, and far more common, there was extermination at a distance, in which the extent of the killing was so vast that the killer tended to think in terms of areas on a map rather than individuals. In neither case was the victim perceived as a person, but in the first case the killer at least saw the immediate consequences of his act, whereas in the second case he did not. The "close-range" killers in Vietnam were confronting *something*, even if it had little to do with the root dissatisfactions in their lives (one of the Huey pilots in Harvey's book re-enlisted because he couldn't tolerate the demands of civilian life).

But for the "long-range" killers—which in a sense includes all of us—do we need any explanation at all? Governments have always tried to keep their soldiers from thinking of "the enemy" as human by portraying them as monsters and by preventing contact ("fraternization"). Modern weaponry makes it easy for anyone to be a mass killer without much guilt or stress. Flying in a plane far above an impersonally defined target and pressing some buttons to turn fifty square miles into a sea of flame, is less traumatic to the average middle-class American than inflicting a superficial bayonet wound on a single soldier. The flier is protected from intimate contact with the victims of his mutilations. He cannot see the women and children being horribly burned to death they have no meaning to him. Violence-at-a-distance, then, was popular just because it was so easy—another expression of the Toilet Assumption. It could be placed in the same category with the arguments of "right-to-life" adherents, who are very conscious of dead fetuses in hospitals, but assume a bland indifference to the unseen suffering of mothers dying in childbirth, or butchered by quack abortionists; or to the untold misery inflicted by and on those millions of unwanted children.

The Toilet Assumption, however, isn't enough to account for our love of violence-at-a-distance. Not everyone who has a gun uses it, and not everyone who has an atom bomb drops it. One must explain why the United States has developed more elaborate and grotesque techniques for exterminating people at a distance than any nation in history. Our preference for slaughter from the air certainly has some practical basis in the need to insulate carefully reared soldiers from the horrors they cause, but practical considerations alone hardly account for the fiendishness of the weaponry. Can this all result from the miseries and frustrations of American life?

The Familiar Stranger

Perhaps it isn't an accident that Americans are so fond of killing from a distance—perhaps the distance itself carries special meaning. Perhaps Americans enjoy the impersonal killing of people who can't fight back because they themselves suffer impersonal injuries from mechanical forces against which they, too, are powerless.

There are indeed two ways in which this occurs. The first comes from our attempts to avoid conflict by increasing personal autonomy, which, as we have seen, only makes the conflicts more indirect. When we create bureaucratic mechanisms to avoid conflict with our neighbors, we just find ourselves beleaguered by some impersonal far-off agency. When you fight with your neighbor you can yell at each other and feel some relief, perhaps even make it up or find a solution. But there's little satisfaction in yelling at a traffic jam, or a faulty telephone connection, or an erroneous IBM card, or any of the thousand petty and not so petty irritations to which Americans are daily subjected. Most of these irritations come from vast, impersonal institutions with which the people we encounter are only vaguely connected ("I only work here"). We feel helpless before the size and complexity of these institutions, and trying to find out who's responsible for our difficulties is such an overwhelming task that it is often abandoned even by middle-class persons. (The poor seldom try.) Comedians have become adroit at satirizing this situation, but aside from laughter and vague expressions of futile exasperation at "the system" we can do little to relieve our feelings. The effort needed to ward off even the most obvious forms of commercial exploitation in our society would prevent us from leading a normal life. Like Looking-Glass Country, it takes all the running you can do just to stay in the same place.

Powerlessness has always been the common lot of most people, but in a preindustrial age you could at least locate the source of injury. If a nobleman beat, robbed, or raped you, you could at least hate the nobleman. If a hospital removes a kidney instead of an appendix, or when there is only one kidney to remove, whom can we hate? The orderly who brought the wrong record? The doctor who failed to notice? The poor filing system?

The more we try to solve our problems by increasing personal autonomy, the more we find ourselves at the mercy of these mysterious, impersonal, and remote mechanisms that we have ourselves created. Their indifference is a reflection of our own.

Our fondness for violence at a distance is both an expression of, and a revenge against, this indifference. We sent bombers to destroy "communism" in Vietnam to avoid meeting our needs for cooperation at home. But some of that savagery was also aroused by the remoteness itself. Distant and unknown enemies have a special meaning for us—we associate them with the unknown forces that beset us. In other words, the very fact of Vietnam's remoteness and strangeness increased our hatred and our willingness to use sadistic and genocidal instruments. This becomes clear when we compare Vietnam to Cuba: both are small countries involving no real threat to our power, but one is near and familiar, while the other is far away, with an oriental population. We would never have used in Cuba the instruments of mass destruction we employed in Vietnam.

The second way we suffer from impersonal forces has to do with our opposing needs for stability and change. All societies must allow for both, since adaptation to the environment requires both new responses and the consolidation of old ones, while our personal happiness demands both familiarity and novelty in everyday life. Every society develops some way of realizing these contradictory needs.

Our society handles the problem by giving completely free rein to technological change and opposing formidable obstacles to social change. Since, however, technological change forces social changes upon us, this has had the effect of abdicating all control over our social environment to a kind of whimsical deity. While we think of ourselves as a people of change and progress, masters of our environment and our fate, we are no more so than the most superstitious savage, for our relation to change is entirely passive. We poke our noses out the door each day and wonder breathlessly what new disruptions technology has in store for us. We talk of technology as the servant of humanity, but it is a servant that now dominates the household, too powerful to fire, and upon whom everyone is helplessly dependent. We tiptoe about and speculate on his mood. What will be the effects of such-and-such an invention? How will it change our daily lives? We never ask, "Do we *want* this, is it worth it?" (We didn't ask ourselves, for example, if the trivial convenience of the automobile would really offset the calamitous depersonalization of our lives that it brought about.) We simply say, "You can't stop progress," and shuffle back inside.

We pride ourselves on being a democracy but we are in fact slaves. We submit to an absolute ruler whose edicts and whims we never question. We watch him carefully, hanging on his every word; for technology is a harsh and capricious king, demanding prompt and absolute obedience. We laugh at the Luddites who went around smashing machines in early nineteenth-century England, but they at least confronted the issue. Americans for the past 150 years have passively surrendered to every degradation our technological ingenuity has brought about. We laugh at the old lady who holds off the highway builders with a shotgun, but we laugh because we're Uncle Toms. We try to outdo each other in singing the praises of the oppressor, although the value of technology in increasing human satisfaction remains at best undemonstrated. We say this or that invention is worthwhile be-

cause it generates other inventions—because it is a means to
some other means, not because it achieves any ultimate human
goal. We play down the "side effects" that so often become the
main effects and completely negate the alleged benefits. The advantages of *all* technological "progress" will, after all, be totally
nullified the moment nuclear war breaks out (an event which,
given the number of fanatical fingers close to the trigger, is only
a matter of time).

I don't believe in the noble savage, and I'm not pushing some
kind of bucolic romanticism. I don't want to put an end to machines, I only want to end their mastery, to restore human beings
to full equality ind initiative. As a human I must protest that
being able to sing and eat watermelon all day is no compensation
for being beaten, degraded, and slaughtered at random, and this
is our current relationship to our technological order.

Not all these ills can be blamed on capitalism. The Soviet
Union and other planned economies are as enslaved as we. Technology makes core policy in every industrialized nation, and the
humans adjust as best they can.

The much-vaunted "freedom" of American life is thus an illusion. We are free to do only what we are told, and we are "told"
not by a human master but by a mechanical construction.

But how can we be the slaves of technology—isn't technology
just an extension of ourselves? The metaphor is as misleading as
it is self-congratulatory. The forces to which we submit so abjectly were created not by ourselves but by our ancestors. What
we create will in turn rule posterity. It takes time for the social
effects of technological change to make their appearance, and by
then a generation has usually passed.

Science-fiction writers have long been fascinated with the idea
of being able to create material objects just by imagining them,
but actually it's just an exaggeration of what normally takes place.
Technology is materialized fantasy. We are ruled today by the
materialized fantasies of previous generations.

The Vengeance of the Dead

This is why the idea of the tyrannical father has never disappeared from American culture. For although in everyday family

life the despotic patriarch is a rare curiosity, the *idea* is familiar to every American. Wolfenstein and Leites[9] observe, for example, that in early American films the hero's father would be portrayed as a kindly, bumbling, ineffectual figure; but the hero would always come into conflict with another male authority—a cattle baron, tycoon, political leader, or racketeer who was powerful, evil, despotic, iron-willed, and aggressive, quite unlike the kindly old father.

I can't help feeling that this portrait is drawn from some reality, and I suspect that it has something to do with technology. We certainly treat technology as if it were a fierce patriarch we're deferential, submissive, alert to its demands. We feel spasms of hatred toward it, and continually make fun of it, but do little to challenge its rule. Furthermore, since the technological environment that rules, frustrates, and manipulates us is a materialization of the wishes of our forefathers, it's quite reasonable to say that technology *is* an authoritarian father in our society. The American father can be a good-natured slob in the home precisely because he's so ruthless toward the nonhuman environment—leveling, uprooting, filling in, building up, tearing down, blowing up, tunneling under. This ruthlessness affects his children only indirectly, as the deranged environment afflicts the eyes, ears, nose, and nervous system of the next generation. But it affects them nonetheless. Through this impersonal intermediary we inflict our will upon our children, and punish them for our generous indulgence—our child-oriented, self-sacrificing behavior. It's small wonder that the myth of the punitive patriarch stays alive.

In a way this is just another example of the first process we described: the attempt to avoid conflict by creating a false illusion of autonomy—imagining that because we put mechanical compartments between people we've created a self-governing paradise. It's a kind of savage joke. We say: "Look, I'm an easygoing, goodnatured, affectionate father. I behave in a democratic manner and treat you like a person, never pulling rank. As to all those roads and wires and machines and bombs and complex bureaucratic institutions out there, don't concern yourself about them that's my department." When the son grows up he discov-

ers the fraud. He learns that he's a slave to his father's unconscious and unplanned whims—that the withheld power was crucial. But by this time he has also learned the system of avoiding
conflict through impersonal mechanisms and is ready to inflict
the same deception on his own children.

Margaret Mead describes a mild, peaceful tribe—the Arapesh—who share this device with us.[10] Whenever an Arapesh
male is angry at one of his neighbors, he never attacks him directly. Instead, he gets some of his "dirt" (body excretions, food
leavings, and so forth) and gives it to a sorcerer from another
tribe nearby. The sorcerer may or may not use it to destroy the
victim through magic. If the man dies years later, his death will
be attributed to this sorcery, though the quarrel be long since
forgotten. The Arapesh see themselves as incapable of killing
each other—they don't even know any black magic. Death
comes from the foreign sorcerers.

Our enlightened civilization uses precisely the same model.
We love and indulge our children and would never dream of
hurting them. If they are poisoned, bombed, gassed, burned, or
whatever, it's surely not our fault, since we don't even know how
to manipulate these objects. The danger comes from outside.
Perhaps long ago we did something to deliver them into these
impersonal hands, but we've forgotten, and in any case it isn't
our responsibility. Technology, in other words, is our foreign sorcerer.

The joy of killing, then, is strengthened for Americans if the
"enemy" is distant and impersonal. Since injury comes to us from
remote sources we must find a remote victim on which to wreak
our vengeance.

Since the "real" enemy is our technologically strangled environment (created by ourselves and our ancestors), it may seem
ironic that we avenged ourselves by killing poor people who
never experienced this kind of environment before we inflicted
it upon them. We used what oppresses us to oppress other human beings. But this has always been true of downtrodden
classes—afraid to attack the oppressor they take out their rage
on each other. Black men for centuries squandered their rebellion in fraternal slaughter, and we, enslaved by our Franken-

stein-monsters, behave no differently. It isn't likely that Americans will ever join together and consign their weapons, poisons, and other life-hating implements to oblivion. Misery loves company more than its own end. And Americans love their machines more than life itself—more even than their children, toward whom, as we shall see in the next chapter, feelings of adoration and resentment are in precarious balance.

THREE

Women and Children First

Does she put the kids to bed, then read or watch TV
While you're out being someone that you want to be?

HOLLY NEAR

One of the film classics of the sixties was Mike Nichols's *The Graduate,* an odd mixture of satire and sentimentality. It attacked some of the most painful ailments of American society, while giving resounding approval to the cultural addiction on which they depend—our unquestioning faith in romantic love. The satire was fresh, imbued with the energy and spirit of the sixties, but the romanticism was in the oldest Hollywood tradition: the American dream of love triumphing over social reality. Yet it was a reformed and purified version of that dream, and this gave it some novel twists.

The climax of the film was the interruption of a wedding ceremony. This has always been a popular theme in American films, for it dramatizes the everlasting conflict between social forms and human feelings. Films like *The Graduate* are a kind of ritual, celebrating the power of such feelings and the conviction that they should occasionally win.

But in earlier films the conflict was usually confined to the age-old Hollywood question of whether to choose the more romantic and less conventional of two prospective mates. The stop-the-wedding scene was typically comic, with little disruption of the ceremony. *The Graduate* was new in that the hero made no attempt to cover or mask his feelings, so that the ceremony was brutally and irretrievably shattered—the hero forced to do physical battle with the participants. In this scene the old theme was

presented with such baldness that it seemed new and revolu
tionary.

For any generation born before World War II, rituals, cere-
monies, and social institutions have an inherent validity that
makes them intimidating—a validity that has priority over human feelings. One would hesitate to disrupt a serious social occasion for even the most acute and fateful need, unless it could
be justified in social rather than personal terms. Doris Lessing
and Shelley Berman have both observed (in the case of people
confronted with aircraft whose integrity has been cast in doubt)
that most people would die quietly rather than make a scene.

Many younger people no longer share this allegiance. They
don't see social occasions as having automatic validity—social
formality is deferred to only when human concerns aren't pressing. Stoicism is not valued. Thirty years ago, on the other hand,
a well-brought-up young man like the hero of *The Graduate*
would have stood passively watching while his personal disaster
took place. Cinematic comedy often made use of this meek deference—think of the cops-and-robbers chases in which both participants would briefly interrupt their frantic efforts in order
to stand at attention while the flag or a funeral procession
passed by.

This change was responsible both for the character of radical
protest in the sixties and for the angry responses of older people
to it. Sitting-in at a segregated restaurant, occupying a campus
building, lying down in front of vehicles, pouring blood in office
files—all depended heavily on a willingness to make a scene and
not be intimidated by a social milieu. And this was precisely
what so enraged older people. They were shocked not so much
by the radicalism of young people as by their bad form. That
students could be rude to a public figure was more shocking to
parents than that the public figure was sending their children to
their deaths in an evil cause.

Yet the change was one that the parents themselves had created, for it was based on child-centered family patterns. While
Europeans have always felt that American parents gave far too
much weight to their children's needs and far too little to the

demands of adult social occasions, Dr. Spock's emphasis on allowing the child to develop according to her own potential carried the trend even further. It focused the parents' attention on the child as a future adult, who could be more or less intelligent, creative, and healthy according to how the parents behaved toward her. This was unlike the older view that the child had a fixed personality to which the parents tried to give a socially acceptable wrapping. The old method was based on the military model: you take people who are all different and get them to behave outwardly in a uniform manner, whether they're inwardly committed to this behavior or not. Thus there was a sharp distinction between one's outer and inner worlds. The child or recruit was *expected* to harbor inner feelings of rebellion or contempt, so long as these were not expressed outwardly.

The new method gives much more responsibility to the parents, who must now concern themselves with the child's inner state. They are no longer trying merely to make the child well-behaved—for them personality is not a given, but something they can mold. The parents under the old method felt they had done their job well if the child was obedient, even if he turned out dull, unimaginative, surly, sadistic, and sexually incapacitated. Spockian parents feel it's their responsibility to make their child into the most all-around perfect adult possible, and although what this leads to may look like "permissiveness," it's actually more totalitarian, for the child no longer has a private sphere. His entire being has been taken over by parental aspirations: what he is *not* permitted to do is take his own personality for granted.

Under the old system, for example, the parents would feel called upon to chastise a child defined as bright but lazy, and if they forced him to spend a fixed amount of time staring at a book—whether he learned anything or lost all interest in learning—they would feel justified and relieved of all moral responsibility for him ("I don't know why he's so bad, I beat him every day."). Today parents feel required not just to make him put in time but to "motivate" him to learn.

The tradeoff for having her whole personality up for grabs is that the child's needs are paid much more attention. The old method demanded that these needs be subordinated to social reality: in the most casual social encounter the parents would be willing to sacrifice the child's sense of truth and fair play ("kiss the nice lady"), her bodily needs ("you'll just have to wait"), and even parental loyalty ("she's always stupid and shy with strangers"). For the parent who loves her, to throw her to the dogs for something so trivial as etiquette makes a deep impression on the child. She sees the parent nurturant and protective in situations that seem much more important and dangerous, so why not here? Since she can't *see* anything important enough to justify this betrayal, all social situations tend to acquire a sacred, intimidating air. When the parents put this mysterious situation above all else, it acquires the same importance for the child.

But Spock-taught parents, fired with the goal of molding the child's total character, were much less inclined to sacrifice her to the etiquette concerns of strangers—the artist working on a masterpiece doesn't let guests use it to wipe their feet on. As a result, their children have grown up feeling that human needs have validity of their own. Social occasions are less sacred to them than they were to earlier generations.

Hence the hero of *The Graduate* is not intimidated by the wedding ceremony but wails out his pain, and the heroine, until then bewitched by social forms, is disenchanted, rescued, and redeemed. But what of the parents, who have given their children the power to confront what they are unable to resist themselves? How do they react? In *The Graduate* they show vindictive hatred, and this also was a new departure, for in the older films the custodians of social forms were merely left openmouthed, or slyly smiling (secretly glad), or futilely shaking their fists. But here they attack viciously and a melee ensues. The hero, grabbing a large cross from the altar, beats the mob off with it, and then uses it to bar the church door from the outside, permitting the couple to escape.

The wielding of the cross exposed a peculiarity of contemporary parent-child relationships. As every moviegoer knows, one carries a cross to ward off vampires, and putting a cross on a door prevents the vampires from getting through. In *The Graduate*, as in upper-middle-class America generally, the parents relate to their children in a somewhat vampiresque way. They feed on the child's accomplishments, sucking sustenance for their pale lives from vicarious enjoyment of his or her development. At the beginning of the film, for example, at a party given to celebrate the hero's return from an honor-laden college career, family and friends clutch and paw him like a valuable artifact. And later, when he models some diving equipment in the family pool, the hero becomes a mannikin on which his father displays his affluence to his friends.

In a sense this sucking is appropriate since the parents give so much—lavish so much care, love, thoughtfulness, and self-sacrifice on their blood bank. But this is little comfort for the child, who at some point must rise above his guilt and live his own life—the culture demands it of him. And after all, a vampire is a vampire.

Spock's Impact

It would be unfair either to credit or to blame Dr. Spock for changes in the American character. His books on child-rearing would not have been so popular and influential had they not been firmly rooted in existing American values and attitudes. At the same time, however, they strengthened and nourished those attitudes. In particular, they reinforced three trends in American family and child-rearing patterns: permissiveness, individualism, and feminine domesticity. The first two have been with us for at least two centuries, but the last was a post-World War II phenomenon—a twenty-year interruption of an older trend in the opposite direction.

Often it's assumed that permissiveness in child-rearing is a recent American development, but this is clearly not the case. While every generation of Americans since the first landing has imagined itself to be more permissive than the previous one, foreign visitors have refused to notice any ups or downs in the unremitting stream of American laxity: They have stoutly and consistently maintained since the seventeenth century that American children were monstrously undisciplined.[1] Spock, in any case, has always emphasized the child's need for parental control and the importance of not letting him become a tyrant in the home. The areas in which he reinforced "permissiveness" had to do not with social behavior but with such matters as feeding schedules and toilet training, and even here he merely revived practices current in America and England prior to the middle of the eighteenth century. While Spock has become a *symbol* of permissiveness in child-rearing, I think we'll learn more about his impact by looking at the other two patterns: individualism and feminine domesticity.

Spock's work is in the old American tradition that every individual is unique and has a "potential." This potential is viewed as innate, partially hidden, gradually unfolding, and malleable.[2] The parent cannot simply coerce the child into a set pattern of behavior because it's important to our achievement ethic that a child realize her maximum potential, and that means taking into account real or imagined characteristics of her own. The parent is given not clay but some more differentiated substance with which to mold an adult.

Spock is concerned about what he feels to be our excessive child-centeredness, but he sees no escape from it: "I doubt that Americans will ever want their children's ambitions to be subordinated to the wishes of the family or the needs of our country."[3] He suggests that children would be happier if parents would stick to whatever principled guns they have, but this hardly balances the general thrust of his work. From the very beginning Spock's books have encouraged Pygmalionesque fantasies in mothers—stressing the complexity and importance of the task of creating a person out of an infant. His good sense, tolerance, humanity, and uncanny ability to anticipate the anxieties that everyday child-rearing experiences arouse in young mothers seduce them into accepting the challenge. Deep in their hearts most middle-class, Spock-taught mothers believe that if they did their job well enough, all their children would be creative, intelligent, kind, generous, happy, brave, spontaneous, and good—each, of course, in his or her own special way.

It was this challenge and this responsibility that led mothers to accept the third pattern that Spock reinforced—feminine domesticity. For until quite recently, when he finally bowed to the demands of feminists and acknowledged the legitimacy of their resentments, he has always maintained that a woman's place is in the home. He emphasized the importance and the difficulty of the task of child-rearing and gave it priority over all other possible activities. He suggested government allowances for mothers compelled to work on the grounds that it "would save money in the end"—implying that only a full-time mother could avoid bringing up a child who was a social problem. He allowed reluctantly that "a few mothers, particularly those with professional training," might be so unhappy if they didn't work that it would affect the children—the professional training was seen as a kind of unfortunate accident that could no longer be undone. The mother had to feel "strongly" about it and have an "ideal arrangement" for child care. Otherwise Spock tried to induce guilt: "If a mother realizes clearly how vital this kind of care is to a small child, it may make it easier for her to decide that the extra money she might earn, or the satisfaction she might receive from an outside job, is not so important after all."[4]

American women have always had a reputation for independence—Tocqueville commented upon it in 1830. Our culture as a whole tends to exert a certain pressure for sexual equality, and American women in the nineteenth century were not as protected as women in Europe. In frontier settings they were too important to yield much power or pay much deference to husbands, and among immigrant groups they were often more employable than their husbands. During the present century labor-saving devices reduced the demands of the home to a minimum, education for women increased, women obtained the vote, and contraception undermined the double standard. The direction of events seemed clear.

After World War II, however, a strange thing happened. Although more women were working than ever before, this was not true in the professions. Despite more women going to college, a smaller percentage were using this education in any way. In short, while single middle-class women were becoming more and more liberated, married middle-class women were embracing a more totally domestic existence than ever before. But how was this achieved? How could educated women devote their entire lives to a task so shrunken? How could they make it fill the day, let alone fill their minds? To some extent Parkinson's Law ("work expands to fill the time available to complete it") applies to such situations, especially with the aid of the advertising industry, which continually invents new make-work chores and new standards of domestic perfection.

But the main factor in the domestication of the middle-class American woman was the magnification of the child-rearing role. Child-rearing is not a full-time job at any age in and of itself. In every other society throughout history women have been busy with other tasks, and reared their children as a kind of parallel activity. The idea of devoting the better part of one's day to child care seldom occurred to anyone because few women ever had time for it before, and when they did, they usually turned the job over to a servant. Occasionally someone fiercely determined to produce a genius would spend hours a day trying to teach an infant Latin or Greek, but these were eccentricities. In our society it is as if every middle-class parent were determined to rear a John Stuart Mill; it turns one a bit queasy to see them walking about with signs on them so their three-year-olds will learn to read, or complaining that their children are not learning enough in nursery school.

This is not to say that child care *cannot* fill a day. The modern suburban home is neither built nor equipped in a way that allows for the comfortable or healthy management of an eighteen-month-old child. Living in the suburbs also forces the mother to be constantly driving her children about from one activity to another. Anyone could add to the list of anomalies created by our being a child-oriented society in the face of a technological environment that is antagonistic to children. One has only to see a village community in which women work and socialize in groups with children playing nearby, also in groups—the older children supervising the younger ones—to realize what's awkward about the domestic role in America. Because the American mother is isolated, she can engage in only one of these three activities (work, socializing, child-rearing) at a time, with effort, two— hardly a satisfying occupation for a civilized woman.

But most important, the American mother has been told: "You have the capacity to rear a genius, a masterpiece. This is the most important thing you can do, and it should rightfully absorb all of your time and energy." With such an attitude it's easy to expand child-rearing into a full-time job. For although Spock has many sensible passages about not martyring oneself to one's children ("needless self-sacrifice sours everybody"), the temptation to do so is enormous when there's so little else. In the tedium of domestic chores, this is the only thing important enough to be worthy of attention. We are a product-oriented society, and the American mother has been given the opportunity to turn out a really outstanding product.

Too Much of a Good Thing

Unfortunately, however, there really isn't much she can do to bring this about. At first the child sleeps most of the time, and later spends more and more time playing with other children. It isn't particularly helpful to waken a sleeping infant, and parents aren't very good playmates. The only way she can feel she's putting a proper amount of effort into the task is by cultivating the child's natural entropic tendencies to make more housework for herself; or by upsetting and then comforting the child so she can flex her nurturance and her therapeutic skills. Since she really doesn't know how to create an outstanding adult, and perhaps recognizes, deep in some uncorrupted sanctuary of good sense, that the more actively she seeks it the less likely she is to attain it, the only time she'll feel she's doing her job is when she's meeting minor crises. Naturally this creates a great temptation to induce such crises, indirectly and, of course, without conscious intent.

I once suggested that jovial references to the many roles played by housewives in our society are a way of masking the fact that the housewife is a nobody.[5] A similar effect is achieved by the Story of the Chaotic Day, in which one minor disaster follows hard upon another, or several occur simultaneously (". . . and there I was, the baby in one hand, the phone and doorbell both ringing . . ."). These sagas are enjoyed because they conceal the fundamental vacuity of the housewife's existence.[6] Saying "everything happened at once" is an antidote to the knowledge that nothing ever happens, really.

The emotional and intellectual poverty of the housewife's role is nicely expressed in the universal complaint: "I get to talking baby talk with no one around all day but the children." There are societies in which the domestic role works, but in those societies the housewife is not isolated. She is either part of a large extended family household in which domestic activities are a communal effort, or participates in a tightly knit village community, or both. The idea of imprisoning each woman alone in a small, separate, and self-contained dwelling is a modern invention, dependent on an advanced technology. In Moslem societies, for example, the wife may be a prisoner but at least she is not in solitary confinement. In our society the housewife may move about freely, but since she has nowhere to go and isn't a part of anything anyway, her prison needs no walls.

For a middle-class woman this is in striking contrast to her premarital life. In school she's embedded in an active group life with constant emotional and intellectual stimulation. Marriage typically eliminates this way of life for her, and children deliver the *coup de grace*. Her only significant relationship tends to be with her husband, who is absent most of the day. Most of her social and emotional needs must be satisfied by her children, who are hardly equal to the task. Furthermore, since she's supposed to be molding them into superior beings, she can't lean too heavily on them for her own needs, although she's sorely tempted to do so.

This is the most vulnerable point in the whole system. Even if the American housewife were not a rather deprived person, it would be the height of vanity for anyone to assume that an unformed child could tolerate such massive inputs of one person's personality. In most societies the impact of the mother's character defects is diluted by the presence of many other nurturing agents. In middle-class America the mother not only tends to be the exclusive daytime adult contact of the child, but also has a mission to create a near-perfect being. This means that every maternal quirk, every maternal hang-up, and every maternal deprivation is experienced by the child as heavily amplified noise from which there is no escape.

Maternal Overload

Societies in which deprived mothers turn to their sons for what they cannot obtain from male adults tend to produce men who

are vain, warlike, boastful, competitive, sadistic, and skittish toward women. They have great fear of losing self-control, of becoming dependent on women, of weakness. They often huddle
together in male gangs.[7]

Middle-class American males fit only part of this description
(although American foreign policy is deeply rooted in *machismo*
attitudes). One reason may be that in societies that produce this
kind of male, the mother-son relationship is highly sexualized.
But a seductive mother in a society that gives the child many
caretakers has nothing like the impact she has in a society like
ours, where she is almost the whole world to the child. This
perhaps accounts for the sexlessness of American housewives as
a class. It's as if there were some unconscious recognition of the
fact that even ordinary feminine seductiveness, given the magnification motherhood receives in our society, would be disorganizing to a male child. Since the American mother is omnipres-
ent and so intensely committed to her role, she must be defused,
as it were. Her desexualization is necessary in order not to add
unduly to the already overwhelming maternal input the child
receives.

In many societies a woman is viewed as relatively neuter until
she is married—it's at this point that she becomes a full-fledged
female, a sexual being. Even societies that encourage adolescent
promiscuity sometimes view youthful sexual encounters as just
children playing. In dress, manner, and personal style it's often
the married woman alone who is fully sexual.

In our society the exact opposite is true. Stylistically, it's only
the young unmarried girls who are allowed to be entirely female.
Their appearance is given strong sexual emphasis even before
there's anything to emphasize. But as soon as they're married
they're expected to mute their sexuality somewhat, and when
they become mothers, this neutralization is carried even further.
This means that whatever sexual appeal there is in a malnourished nymphet is made highly explicit, while the kind of mature
and full-blown femininity that has excited European men for
centuries is masked almost beyond recognition. Suburban
housewives in particular often adopt hard, severe, and geometric
hair and clothing styles. The effect seems "masculine," especially

when combined with a bluff, hearty, and sarcastic conversational manner.

It's tempting to see this as compensatory: women cheated of a career express their "masculinity" in the only form left to them. Certainly it seems reasonable to describe as "masculine" a style that imitates the way men in our society behave in all-male groups. And the hair and clothing suggest mobilization—a readiness to participate in some vigorous activity outside the home (Chinese peasant women on the way to the factory seem relaxed by comparison). Such women, however, are typically hostile to women's liberation.

But what's "masculine" and what's "feminine"? Modern psychoanalytic books are full of absurd statements based on the assumption that sex roles in our society embody biological universals. We know by now that there are few characteristics defined in every culture as masculine or feminine. In some societies women are assumed to be stronger, and carry all the heavy burdens. In some societies women are supposed to be impractical and intuitive; in others men are. In most societies women are seen as earthy, men as spiritual; but Victorian England reversed this order. Even within our own society there are odd contradictions: activity is seen as a masculine characteristic, passivity as feminine. Yet men are supposed to move and talk slowly, while women are expected to be active and birdlike in body movement—moving their hands, using many more facial muscles, talking rapidly. A man who is active in the most basic sense of using many muscles from moment to moment is considered "effeminate."

Sex Stereotypes

It should be emphasized, then, that when we talk of "masculine" and "feminine" we just mean the ways these are usually defined in our culture, and since sex role definitions change from time to time, there is ample room for confusion. If women behave in ways that seem to imitate men, we call this masculine; but if customs change, and certain activities get redefined as appropriate for women, are they "masculine" for doing them? Suddenly we realize we've stumbled on a powerful weapon for "keeping women in their place." It's really a very old and familiar weapon, used with great effect against minority groups. It begins with a stereotype—"women can't think logically," for example. If a woman then shows a capacity for logical thought, she's stigmatized as "masculine."

Women are discouraged from professional careers in the same way, and this has a particularly nasty side-effect on the medical profession. To show that it's really "a man's job," the nurturing, helping, healing aspects must be de-emphasized. To get into medical school you must show an aptitude, not for healing but for chemistry, and in courses that are often so ruthlessly competitive (cheating, sabotage, and undermining fellow students are common) that many humanitarian students are disillusioned and turned off. The crucial pre-med course at Harvard, for example, was for years taught by the inventor of napalm. Thus the recruitment of physicians selectively favors cold, ungiving, exploitative, competitive, and mercenary personality types, with a result familiar to all.

Blacks were the first minority group clever enough to invent a solution to this ruse. Instead of trying to escape the black stereotype and become "white," militant blacks accepted the stereotype and said, "black, even stereotyped, is *better* than white." Since white American culture was being strangulated by its alienation from the body, this meant blacks could view themselves as saviors of the society. Women are in much the same position, since alienation from the body and from the emotional life is largely a white male invention. Every negative stereotype can be viewed as a virtue from another angle, and any quality that a society puts down is obviously going to be in short supply, and hence needed to make the society whole and healthy. A curse, in other words, is just a blessing whose time has not yet come.

Like all oppressed groups moving toward liberation, women have been torn between beefing up the parts of themselves squelched by men (assertiveness, achievement, power) and affirming the value of what men think women have too much of (emotionality, intuitiveness, vulnerability). The demand of some feminists for total separation from men is based on the simple fact that it's virtually impossible to do both these things in the face of daily male stereotyping, yet to slight either is intolerable. Black separatism was a response to the same dilemma.

A good example of the dilemma can be found in the field of psychotherapy. Women have been turning away from male therapists and counselors in droves, tired of being defined and manipulated by male anxieties, male preoccupations, male needs for status and control. In some cities, women with psychological skills have formed collectives to provide counseling and therapy for other women. Yet, oddly enough, many feminist therapists begin by aping the style of the most traditional male therapists— adopting the very behavior that led them to reject males as therapists: the cool, remote, controlled, and controlling pose that clings to professional status, hides behind a mask of invulnerability, loves to talk about and analyze feelings but is afraid of expressing them. It often takes some time and a lot of support from more thoroughly liberated women before feminist therapists can permit themselves the kind of open, unpretentious, vulnerable, intuitive, feelingful, natural, informal, action-oriented, and emotionally courageous behavior that comes more easily to women than to men and makes them far more successful therapists. But the dilemma is easy to see: male fear of emotionalty makes them poor therapists, and their cerebral, pedantic training usually cripples them further; yet male power has allowed them to define what therapy is, and when women try to *replace* men, they're tempted to imitate what they're replacing. Insofar as they're still allowing men to define things for them, they will tend to feel that a truly liberated therapeutic style is "just something that women do, not therapy." Yet virtually every important therapeutic innovation of the last fifteen years has been an assault on the traditional male psychiatrist's gimmick of hiding emotional timidity behind a professional analytic mask.

The same problem comes up around careers in general. Whenever people talk about women and careers, the objection is always made that many women don't *want* careers. Often women agree with this statement, but with the puzzled uneasiness that people always feel when obliged to accept a formulation that makes them lose either way. The problem once again is that "career" is in itself a male concept—designed by and for males in our society. When we say "career," it suggests a demanding, rigorous, preordained life pattern to whose goals everything else is ruthlessly subordinated—everything pleasurable, human, emotional, bodily, frivolous. It's a stern, Calvinistic word, which is why it always sounds humorous when applied to occupations of a less puritanical sort. So when a man asks a woman if she wants a career, it's intimidating. He's saying, are you willing to suppress half of your being as I am, neglect your family as I do, exploit personal relationships as I do, and renounce all personal spontaneity as I do? Naturally, she shudders a bit and shuffles back to the broom closet. She even feels a little sorry for him, and bewails the unkind fate that has forced him against his will to become such a despicable person.

A more revolutionary and confronting response would be to admit that a "career," thus defined, is indeed undesirable—that (now that you mention it) it seems like a pernicious activity for *any* human being to engage in and should be shunned by both men *and* women. No, she doesn't want a "career," nor do most humans, with the exception of a few males crazed by childhood deprivation or Oedipal titillation with insatiable desires for fame, power, or wealth. What she wants is meaningful and stimulating activity, excitement, challenge, social satisfactions—all the things that middle-class men and women get from their jobs whether they're defined as "careers" or not. But she's rarely willing to pay the price that narcissism seduces men into paying, and therefore accepts the definition of herself as the inferior sex. The revolutionary stance would be to say: "My unwillingness to sacrifice human values to my personal narcissism and self-aggrandizement makes me the superior sex." Such a stance would liberate both sexes: women would be freed from their suffocating domestic stagnation, and men would be liberated from their enslavement to the empty promise (just out of reach, and unsatisfying even when grasped) of "success." Both could then live in a gratifying present, instead of an illusory future and an ill-remembered past.

Women have long been stereotyped as bastions of conservatism—a stereotype supported by attitude surveys. Even war, the

most absurd and vicious of all the games men play, has rarely
produced a feminine revolt.

In our society, however, feminine conservatism is a role into
which women are inducted by men. Having created a technological juggernaut by which they're buffeted daily, men tend to use
their wives as opiates to soften the impact of the forces they have
set in motion against themselves. Consider suburban living: husbands go to the city and participate in the twentieth century,
while their wives are assigned the hopeless task of trying to act
out a pathetic bucolic fantasy. In their jobs the husbands must
accept change—even welcome and foster it—however threatening and disruptive it may seem. They don't know how to abstain from colluding daily in their own obsolescence, and they're
frightened. Such men tend to make of their wives an island of
stability in a sea of change. The wife becomes a kind of memento,
like the bit of earth the immigrant brings from the old country
and puts under his bed. He subtly encourages her to espouse
absurdly old-fashioned views which he then ridicules when he's
with his male associates. There is a special tone of good-natured
condescension with which married men gathered together discuss the conservatism of their wives, and one senses how elegantly the couple's ambivalences have been apportioned between them. ("It's a great opportunity for me, but of course the
wife doesn't like to move—she has ties in the community, and of
course the children in school and all....") It permits the husband
to be far more adaptable and amenable to change than he really
feels.

But ultimately this kind of emotional division of labor always
backfires. Freed from the necessity of confronting his own resistance to change, and having insulated his wife from the exciting
and enjoyable aspects of novelty, he tends to become bored with
her and somewhat lonely, and soon seeks greener pastures. This
likelihood is increased if the wife has desexualized herself in the
service of motherhood, but the housewife role pretty much
dooms her anyway. Our society is founded on overstimulation—
on the creation of complex desires that can't directly be gratified,
but that seduce people into a lot of striving and buying in the
vain effort to satisfy them. Most of these desires are vaguely or

blatantly sexual—erotic delights are attached by advertisers to most of the goods and services that can be bought in the United States. The goal of American commerce, in other words, is to arouse kinky needs that defy satisfaction—in this way an infinite number of products can be inserted in the resulting gap. In such a culture availability itself is a turn-off.

But what of the wife? Is she merely passive in all this? What of *her* boredom and *her* desires? Those women who have been untouched by the women's movement are often simply trapped in their own conservatism and loyalty. They smother their yearnings, ignore their boredom, and muddle along as best they can with the aid of a rich fantasy life. Doing what one feels she *should* do can be very comforting. Only if and when her husband turns to someone else is she likely to question her position—especially if she has a comfortable home and is well off.

More and more women, however, are undermining the myth of feminine conservatism, whether they join the work force or simply change their lifestyle. In fact, the impact women have had on those around them (both men *and* women) has often been greater off the job than on. The first women to enter any occupation tend to behave just like the men—even a little more so. It's only when their numbers approach equality that styles of work begin to be affected, and not always even then. But women who have had their consciousness raised about relationships have often transformed those relationships. Instead of being left behind by working husbands, in many marriages today it's the husbands who are being left behind—puzzled, frightened, and unable to keep pace with the dazzling personal changes their wives are going through.

Yet these women are still a small minority, and while I don't want to ignore the changes that are beginning to happen all over the country, I also don't want to ignore the fact that the majority of men and women are still thinking as they did twenty years ago.

The Playboy Ideal

Men, like all dominant groups, have been pretty successful in getting women (like other "minority" groups) to accept any definition of their character that's been convenient for men. One of the oldest tricks has been to maintain that dominance is sex-linked, so that if a woman isn't submissive, she's accused of being "unfeminine." This is an ingenious way to maintain superior status and has been quite successful. On the other hand, men lose a lot by thus hobbling the peronalities of their mates. Whenever men have succeeded in convincing their wives that some human response was unfeminine," they have sought other women who possessed it.

One has only to think of Sophia Loren or Elizabeth Taylor to realize that dominance and aggressiveness in women detracts little from their sexual attractiveness. On the contrary, women who have been taught too well that aggressiveness is "unlady-like" often seem sexless. There is a depth in the human psyche at which all feelings merge, and the disparagement of one constricts and dampens all the others.

Yet it can't be denied that the female sexual ideal in America is nonaggressive and nonthreatening, to the point of caricature. Take, for example, the film personality of the much-idolized Marilyn Monroe: docile, accommodating, brainless, defenseless, totally uncentered, incapable of taking up for herself or even knowing what she wants or needs. A sexual encounter with such a woman in real life would border on rape—the idea of "consenting adults" wouldn't even apply. The term "perversion" seems more appropriate for this kind of yearning than for homosexuality or bestiality, since it isn't directed toward a complete being. The Marilyn Monroe image was the ideal sex object for the sexually crippled and anxious male: a bland erotic pudding that would never upset his delicate sexual stomach.

It's important to realize that this *Playboy* Ideal is a sign of low, rather than high, sexual energy. It suggests that the sexual flame is so faint and wavering that a whole person would overwhelm and extinguish it. Only a vapid, compliant ninny-fantasy can keep it alive. It's designed for men who don't really like sex but need it desperately for tension-release—men whose libido is mainly wrapped up in achievement or dreams of glory. The Marilyn Monroe image is thus more a capitalistic ideal than an erotic one.

Many women have adapted to this male disability, however, by playing the role of dummy and trying to sneak their own needs and personality in around the edges, but it's obviously a strain. What usually happens is that the part of themselves they protect their husbands from is unleashed on their sons, with the result that the sons grow up as nervous about encountering a whole woman as were the fathers.[8]

In other words, the encounter avoided by the father is simply passed on to the son, with interest. The father may successfully hide from it, but from the society's viewpoint it has only been postponed. Women are going to assert their power either as equals in society and everyday life, or as mothers over their sons privately, unconsciously, vengefully—in ways that are frightening and overwhelming to the adult males of tomorrow. The fact that so many "heterosexual" men in our society can only get it up for a giggling robot is a far more serious social problem than may appear at first glance.

Ambitious, "self-made" men, for example, tend to have weak, passive, or absent fathers, and driving, demanding mothers, who turned to their sons for what their husbands were incapable of giving. This means that a lot of our culture—its institutions, its ideology—was formed by men who felt in danger of being overwhelmed by women.

It shows. And it seems to be deepening. Violence against women, for example, has always been popular in the media— Hitchcock never tires of showing women menaced, raped, or murdered. But the current rage for such violence has reached new levels of absurdity, perhaps because women are taking up for themselves again after decades of supine docility. Whenever men have felt unable to stand up to women, to meet them on equal terms emotionally, they have tried to reduce the threat by social manipulation and control. And when this control shows signs of breaking down, the worship of violence increases.

It's hard to choose any single example, given the fascination with violence in films, TV drama, and the news. My favorite, however, for sheer inanity, is the film *Straw Dogs*, which has been a cult item for S-M freaks for several years. The script reads like an old Charles Atlas advertisement: the 97-pound weakling

appears as mild-mannered mathematician with a sexy, provoca-
tive wife, while the bully of the beach takes the form of assorted
yokels. They rape the wife (a yokel herself, naturally, and not up
to the mental level of the hero, who stares at a blackboard all day
and writes equations), but this fails to rouse him ("she asked for
it"). Nor do their personal insults. When they pursue another
man, however, Superman comes righteously out of the closet and
kills or mutilates the whole gang. It's fundamentally a comic-
book film, made by and for emotional ten-year-olds. Some
people go for the gore, but *Straw Dogs* is fundamentally the revenge fantasy of an impotent little boy unable to cope with his
mother's sexuality.

The women's movement is thus engaged in a double rescue:
the short-run rescue of women from vindictive men and the
long-run rescue of sons from vindictive mothers. The important
thing is to break the cycle, for when correction begins in any one
part of it, the rest will change automatically.

The Exorbitant Gift

Men have usually used the welfare of the child as justification for
the domestication of women, so we can see why these same children might encounter some resentment when they grow up.
Children are not really responsible for the bad bargain the parents have made with each other, but they lend moral credit to it.
Indeed, "for the children" is a kind of notary seal given to all bad
marital bargains. And since the child is the justification for the
parents' unsatisfying life together, they can't help but blame him
when they begin to suffer from it. The husband's ambition and
the wife's domesticity originally promised their own rewards and
didn't need to be buttressed by thoughts of the child's future—
just as a mutually profitable deal between two businessmen
doesn't initially require a written contract. But such a contract
binds them if there is a change of heart, at which point one of
them might say, "If it weren't for the contract, I wouldn't go on
with this." Similarly, as ambition and domesticity fail to bring
happiness to husband and wife, both begin to say, "If it weren't
for the children, I might chuck this and do something more interesting (enjoyable, fulfilling, exciting, relaxing)." But you can't

tear up a child like a contract, or even admit to wanting to be
without it. This means the child is not only a scapegoat, but a
scapegoat that can't be attacked. The result is often a vague resentment toward youth—a resentment with roots in the parents'
discontent with their own lives. It's a condition ideally suited to
produce anger toward young people who live differently and
more pleasurably than did the parental generation—hence the
youthhatred of the sixties.

This is not to say that parents don't make heavy sacrifices for
their children in a child-oriented society like ours. I'm trying to
explain why these sacrifices are resented. Parents in many societies make severe sacrifices for their children, but they usually
"pay off" in some way, or at least lead to some predictable outcome. In our society parents never know exactly what their sacrifices will lead to, although they have many fantasies about it.

In the recent past, for example, and in working-class families
today, parents sacrificed in order to prepare their children to be
economically and socially better off than they were, and often
hated their children for fulfilling this goal and leaving them behind. Now middleclass parents sacrifice in order to prepare their
children to be emotionally better off—more expressive, carefree, creative, relaxed, open—and once again resent it when
they succeed. In both cases the parents feel left out of the
triumphs they made possible, and the children feel ashamed of
the parents who wanted them to be superior. The parents never
seem to recognize that success means moving into a new world
from which the parents are automatically excluded. The earlier
parents wanted their children to become rich and respectable
and still remain somehow part of the working-class milieu. The
later group want their children to be less money-grubbing and
more spontaneous, yet still somehow willing to remain on the
same treadmill with the parents.

It isn't surprising that there should be tension between the
generations in a society devoted to change. What's odd is that
each generation, in one way or another, tries to disallow the sexuality of the other. Now many societies try to restrict sexual behavior in the young—our society is peculiar only in that it does
so while simultaneously maximizing their sexual interest and ap-

peal. Meanwhile, those to whom sexuality is freely allowed are desexualized. In other words, there's a severe dissociation between sexual availability and sexual interest in the norms of our society. The fact that the young no longer adhere to these sexual norms arouses an anger in their parents which probably owes something to the parents' own stylistic desexualization. Portrayals of the middle-aged in films and television have traditionally catered to the mental level of school children, who can't imagine their parents making love. When sexuality in the middle-aged or elderly does occur, it's usually in a comic context.

If sexual norms have loosened in the United States, then, our national preoccupation with sexuality has certainly not diminished, while inhibition and prudery have not so much disappeared as donned new guises. Why is there so much preoccupation in America with sexual stimulation, so much tension about the control of gratification—who gets it and how?

Before turning to this question (perhaps by way of introducing it), I'd like to offer one more comment on *The Graduate*. I spoke of what was new in the film, but said little about what was traditional: the belief that happiness and complete fulfillment can be found in a single relationship, that one person can satisfy all needs. Mr. Right meets Mrs. Right and they live happily ever after. It's an Oedipal fantasy, fundamentally, this search for the one-and-only—what real person could satisfy all our contradictory needs and desires simultaneously? Human beings weren't designed to live in isolated couples on desert islands. There's no perfect food that would render all others unnecessary, and there's no perfect person, either. The fact that the hero's new love is the right age needn't fool us—she's as much an Oedipal choice as her mother, Mrs. Robinson, and as sterile. The sequel to *The Graduate*—perhaps to all happy love stories—is *Elvira Madigan*, in which solitary lovers curdle into boredom and suicide.

FOUR

Putting Pleasure to Work

Le diable n'est pas clément
C'est là son moindre défaut
"Que faisez-vous au temps chaud?"
Dit-il à ce vieux diligent.

"Nuit et jour, à tout venant,
Je travaillais, ne vous dèplaise."
"Vous travailliez? J'en suis fort aise:
Eh bien! Jouez maintenant."
(WITH APOLOGIES TO LA FONTAINE)

An English psychologist once found that neurotic anxiety was a good predictor of success and achievement, both for individuals and for nations, thus confirming a long-felt suspicion that something sick forms the driving force for our civilization. Freud argued that "culture . . . obtains a great part of the mental energy it needs by subtracting it from sexuality," and felt civilized people had exchanged happiness for security. He thought it was probably necessary—but was troubled by it: "One is bound to conclude that the whole thing is not worth the effort and that in the end it can only produce a state of things which no individual will be able to bear."[1]

The rising level of anxious and irritable desperation in our society makes it increasingly urgent to evaluate Freud's argument that civilization is a parasite on human eroticism—particularly since the few studies available to us all tend to confirm his hypothesis.[2] Yet one part of his theory is contradicted by the evidence. Freud argued that society "borrows" from sexuality in order to neutralize human aggressiveness, but what evidence there is suggests that restrictions on sexual expression, far from neutralizing aggression, tend to arouse it. Apparently sexual restrictions have some more direct impact on civilization, an impact so powerful that increases in aggressiveness can be tolerated as an unfortunate side effect—or at least have been tolerated until now.

This impact seems to have something to do with energy: "civilized" people are usually described as more energetic or restless than "primitive" ones. This doesn't just mean they *possess* more energy: an improved diet will *enable* some people to work more but won't necessarily implant the desire, while others have quite adequate diets already and yet seem content to lead quiet, carefree lives. I'm talking more about the *utilization* of energy. There seems to be, in other words, some difference in motivation.

Konrad Lorenz once remarked that in all organisms, locomotion is increased by a bad environment. We might then say that sexual restrictions are a way of artificially creating a bad environment, and hence increasing locomotion. Unfortunately, we don't know what *makes* a "bad environment," nor why it increases locomotion. Presumably a bad environment is one that isn't gratifying, and the locomotion is just a quest for more adequate or complete gratification. This makes us think of holding a carrot in front of a donkey on a treadmill, a constant output of energy by the animal depends on the carrot being withheld. Once gratified, the animal would come to a halt, and no further locomotion would take place until the donkey got hungry again.

Hooking Up the Treadmill

Fred Cottrell points out that intermittent energy is relatively useless—solar energy is a good example of the problem. Anything that will translate intermittent energy into constant energy is thus highly valuable from a technological viewpoint.[3] For while a willingness to work constantly is useless without an adequate food supply and other energy sources to exploit, these will also be useless without the willingness to work constantly. A complex society wouldn't be possible, for example, among the Siriono, who lie in their hammocks until impelled by agonizing hunger to hunt for food (at which time they show an energy and endurance far beyond most civilized men).[4] This kind of intermittency was a source of great consternation to early colonial employers, whose work force always melted away on payday, until ways were found to chain them through some system of indebtedness.

But there is a dilemma here. If the donkey never eats, he'll die, but if he does eat, he'll stop working. How can we get a man to work endlessly for a reward which never comes? Obviously we can never solve the problem as long as we're dealing with simple bodily needs that are easily extinguished. A man will work hard for food as long as it's scarce. But what happens when he has a full belly? In order to ensure a steady output of energy we must create some sort of artificial scarcity, for only through such scarcity can an abiding flow of energy be assured.

Sexual desire provides far better raw material than hunger for our project, since it's an impulse that's both powerful and plastic. Its importance becomes apparent when we realize that in nature it's the only form of gratification that is *not* scarce. In fact, it's infinite. This is what people have in mind when they say that sex is the recreation of the poor.

Yet there is no society that doesn't put restrictions on this resource: out of an infinite plenty is created a host of artificial scarcities. It would pay us to look into this matter, since although we live in the most affluent society ever known, its participants seem usually to be in the grip of strong feelings of deprivation, discomfort, and discontent.

The idea of placing restrictions on sexuality was a stunning cultural invention, as important as the acquisition of fire. In it men found a source of energy that was limitless and unflagging— one that enabled them to build empires on earth. By the weird device of making their most plentiful resource scarce, civilized nations have managed to make most of their scarce ones plentiful. On the negative side, however, men have achieved this miracle by making themselves into donkeys, pursuing an inaccessible carrot. They're very elegantly liveried donkeys, it's true, but donkeys all the same. The popular use of the term "treadmill" to describe the institutions through which men make their living shows our dim awareness of this condition.

Trying to find historical beginnings is a futile enterprise. Men are always inventing new follies, most of which are luckily still-born. What we need to explain is why the invention of sexual scarcity has been successful, and not only survived, but grown. Most likely it began with the imposition of restrictions on one group by another: women by men, losers by conquerors.[5] Once begun, it has always had a tendency to spread, for scarcity breeds scarcity just as anger breeds anger. Once people get the idea that there isn't enough of something, they begin to deprive each other of what there is.

Once started, natural selection kept it going. Restless, deprived-feeling tribes tended either to conquer their more contented neighbors or more fully to exploit the resources around them, or both. This cultural advantage was by no means automatic, of course. Without the right kind of environment the restlessness could be merely destructive, and many of the social restrictions were so cumbersome and costly that they absorbed more energy than they made available. But social restrictions have combined often enough with favorable economic conditions to people the planet increasingly with rich scarcity-oriented societies. Occasionally such societies have their sense of deprivation eroded by luxury and are taken over by less "effete," more "virile" (i.e., more deprived-feeling) nations—contributing further to the selection process.

There are many ways to create sexual scarcity, and I want to emphasize strongly that we're not discussing anything as simple as the frequency of sexual encounters or orgasms (although there is evidence that these, too, diminish with civilization).[6] A man can make love as often as he wants and still feel deprived, because his desire has attached itself to someone or something unattainable. The root of sexual dissatisfaction is the ability humans have to generate symbols that can attract and trap portions of their libido. Restrictions on the time, place, mode, and partner do not simply postpone sexual release; they create a permanent deprivation, because humans are able to have fantasies about what they're missing. So while sexual restrictions may intensify the feeling of sexual scarcity, a later relaxation of these restrictions—a trend toward sexual "permissiveness"—may not produce a corresponding *decrease* in the feeling of scarcity. Once you've trained your wolf to prefer cooked meat, you can let it run around the barnyard without any qualms. The best way to generate sexual scarcity is to attach sexual interest to inaccessible, nonexistent, or irrelevant objects; and for this purpose the human ability to symbolize is perfectly designed.

Today this basic technique has become the dominant one. By the time an American boy or girl reaches maturity he or she has so much symbolic baggage attached to the sexual impulse that the mere mutual stimulation of two human bodies seems almost meaningless. Through the mass media everything sexless has been sexualized: automobiles, cigarettes, detergents, clothing. The setting and interpretation of a sexual act come to hold more excitement than the act itself. Thus although the Soviet Union is more puritanical than the United States, there is far more manipulation of the sexual impulse here. Russians are not daily bombarded with bizarre sexual stimuli and deranged erotic associations, so that their simple prudery is ultimately far less repressive.

Across a Crowded Room

Romantic love is one scarcity mechanism that deserves special comment; its main effect, too, is to transform something plentiful into something in short supply. This is done in two ways: first, by having us believe that only one person can satisfy our erotic and affectional desires; and second, by fostering a preference for unrequited, interrupted, or otherwise tragic relationships. Although romantic love always verges on the ridiculous (we would find it comic if a wealthy man died of starvation because he couldn't get any Brussels sprouts), Western peoples generally, and Americans in particular, take it very seriously. Why is this so? Why is love made into an artificially scarce commodity, like diamonds, or "genuine" pearls ("true" love)?

To ask such a question is to answer it. We make things scarce in order to increase their value, which in turn makes people work harder for them. Who would spend their lives working for pleasures that could be obtained any time? Who would work for love, when people give it away? But if we were to make some form of it rare, unattainable, and elusive, and to devalue all other forms, we might conceivably inveigle a few rubes to chase after it.

But this doesn't in itself account for the popularity of romantic love. To see its function doesn't explain it. We can only assume that its strength comes from some intense emotional experience. Few primitive peoples are familiar with it, and it seems to be most highly developed in those cultures in which the parent-child relationship is most exclusive (as opposed to those in which child-rearing is diffused among so many people as to be almost communal).

I like to think of romantic love as a rather glamorous disease like tuberculosis, that often turns ugly in its terminal stages. Its pathology is betrayed by its rigidity: a single act can be lethal to it. This is because it's based on dreams, and dreams are fragile. With people who simply love each other as people, one event is rarely destructive or final or unforgivable. But in romantic love, once can do it, since there's no flexibility when people are trying to force reality into a fantasy. But what *is* this fantasy, and where does it come from?

Since romantic love thrives on absence, one is forced to conclude that it's fundamentally unrelated to the character of the loved one, but derives its meaning from some prior relationship. "Love at first sight" can only be transferred love, since there's nothing else to base it on. Romantic love, in other words, is Oedipal love. It looks backyards, hence its fascination with nostalgia and loss. It's fundamentally incestuous, hence its emphasis on obstacles and nonfulfillment, on tragedy and trespass.[7] Its real goal is not the actual parent, however, but an ideal image that has been retained, ageless and unchanging, in unconscious fantasy.

Romantic love is rare in primitive communities simply because the bond between child and parent is more casual. The child tends to have many caretakers, and to be sensitive to the fact that there are many alternative suppliers of love. The middle-class American child, brought up in a small detached household, usually doesn't have this sense of many options. His emotional life is heavily bound up in a single person, and spreading this involvement over other people as he grows up is more difficult. Americans must make a life task out of what happens effortlessly in many societies. Most Western children succeed in drawing enough money out of their emotional bank to live on, but some always remains tied up in Oedipal fantasy. Middle-class Americans learn early in life that there is one relationship that's more vital than all the others, and they tend both to reproduce this arrangement in later life and to retain, in fantasy, the original loyalty.

The scarcity mechanism on which romantic love is based is thus the intensification of the parent-child relationship. It creates scarcity by a) concentrating one's search for love onto a single person, and b) focusing one's erotic interest on a loved one with whom sexual consummation is forbidden. Making the parent-child bond more intense and more exclusive, *in the face of the incest taboo*, creates erotic scarcity.

We can think of this process as a kind of forced savings: the more we can bind up a person's erotic desires in a tabooed relationship, the less he will seek pleasure in those forms that are readily available. He will consume little and produce much. Savings will increase, profits will be reinvested. So long as he is pursuing what can't be captured, we can relax in the assurance that he will work without cessation into the grave. We have found our donkey.

Chasing Rainbows

I said earlier that bodily gratification is easily obtained, and that in order to get people to strive continually we must introduce restrictions or symbols that will block or filter this gratification and make it incomplete. Hunger, thirst, and sexual desire in pure form can be slaked, but the desire for a body type that was invented by cartoonists cannot. Neither can the desire for fame, power, or wealth. These are invidious needs; they are satisfied only in relation to the deprivation of others. Furthermore, they're purely symbolic and hence have no endpoint. A man hooked on fame or power will never stop striving because there is no way to gratify a desire with a symbol. One cannot eat, drink, or make love with a Nobel Prize, or a presidency, or a controlling interest. One can sometimes purchase bodily gratifications by virtue of such achievement, of course, but they can also be obtained without it. In any case, bodily gratification typically plays a secondary role in the emotional lives of ambitious men—serving as a vacation from or an aid to further productivity.

When we ask why men *do* pursue fame, power, and unlimited wealth so assiduously, the answer is usually that these have become ends in themselves. This is true, but it doesn't answer the question, since the goals have no intrinsic worth. When a means is not used toward an end, but *becomes* an end, then we must assume that the original end has been lost or forgotten. We may have stopped using the carrot, but somewhere in the back of the donkey's head it still exists. He doesn't trot merely because he has come to enjoy the exercise.

When we say of such a man that he is "married to his job" we betray an unconscious understanding of the roots of his striving. Men who pursue these ephemeral goals have most of their emotional funds tied up in the maternal bank. They have a little spending money for daily pleasures, but they are not satisfied with ordinary love. They are committed to an Oedipal fantasy— an emotional long shot that will never pay off. They will work their lives away to achieve a love that is unattainable. They cannot amass enough wealth to buy it, obtain enough power to command it, achieve enough fame to attract it, or do enough good works to deserve it, but still they try. Such men are the most successful donkeys of all.

It is by becoming a donkey that social mobility is achieved. The first class-system that every individual encounters is the division between adult and child, with all the prerogatives, dependencies, and freedoms that go with it. The little boy knows that to replace his father altogether in his mother's affections he must move out of the child-class and into the adult-class, but by the time this happens, the whole fantasy has been relegated to the attic of childhood memory. Yet if the son is in subtle ways encouraged by the mother, because of the father's inadequacies as a provider, or because of special ambitions of her own, he will work out his Oedipal strivings on a social and economic stage. It is this Oedipal fantasy, in fact, that sustains the social climber as he ruthlessly cuts away all the community bonds and loyalties that threaten to hold him down. And it is the value we place on this fantasy game that has made us as a nation so rootless, so community-poor, and so sentimental about motherhood.

An increase in human destructiveness is another outcome. Man may have transformed himself into a donkey, but it's a very irritable donkey. Cross-culturally, the more a society places restrictions on bodily pleasure—particularly in childhood—the more it tends to engage in the glorification of warfare and sadistic practices.[8]

Thus nuclear war holds an unconscious attraction because it offers a final explosive release from the tensions that afflict us, and aggression in less extreme forms also provides an outlet. But war also plays a practical role in maintaining the donkey-carrot syndrome. Our society has become so affluent that it threatens to give the game away—to disclose the absurdity of the scarcity assumptions on which it is based. War creates an artificial scarcity in the economic sphere and thus adds another set of blinders to the donkey's equipment, lest he notice that carrots grow in abundance along the roadside. It is grotesque, for example, that any major service institution in a society as wealthy as ours should experience a financial crisis—*what is our wealth good for if it doesn't provide these services?* Yet in fact all of them—schools, hospitals, universities, local governments, transportation, communications, the arts—are enmeshed in such crises. War, hot or cold, maintains, justifies, and explains these anomalies.

Female readers are perhaps getting irritated by now, since I've been writing about "man" and "his" Oedipal strivings, and using the words "parent" and "mother" interchangeably. The reason is simply that the process I've been describing is largely a male one and does indeed center around the mother-son relationship. The relationship with the mother has always been the primary one for most people of both sexes, and this has important consequences for feminine striving. Ambitious women often have the same kind of intense involvement with their fathers that men have with their mothers, and it's usually more openly acknowledged. But since mothers are so primary, a woman's emotional attachment tends to be more evenly divided between the parents, and this interferes with the kind of monomaniacal investment in career striving that men so often show. This is also why women are in such a good position to liberate our society emotionally. Balance, wholeness, and the ability to blend opposities are more easily available to them. The attack on competitive, egoistic striving that the counterculture began with so much fanfare in the sixties (not without effect, to be sure) will be carried on with more depth, intensity, persistence, and meaning by a new generation of fully liberated women in the eighties, should we be lucky enough to survive that long.

The Futility of Utility

Reactions to the counterculture in the sixties made it quite clear what traditional American culture is all about. One automatic response of older people, for example, to the casual sexuality, clothing, and lifestyle of young people, and to their fascination with altering consciousness through drugs, was to ask, "What is it *for?*" Sometimes various utilitarian motives were imputed: the clothes were to attract attention, the sexual freedom was to produce better marriages (the term "sexual experimentation" captures this utilitarianism nicely), the drugs were to test themselves. The idea that pleasure could be an end in itself was so threatening to the structure of our society that the mere possibility was often denied.

In our society pleasure is allowable only as a means to an end. It must in some way or another yield energy for the economy. Hence the society has developed some special mechanisms for creating scarcity. Pleasure is made scarce, for example, by making it illegal. This makes it expensive and more difficult to obtain, and forces people to compete for what would otherwise be plentiful. Making liquor, drugs, prostitution, pornography, or gambling illegal also opens up new career pathways for the aggressive and ruthless.

Utilitarian assumptions even control our attitudes toward idleness. In public places one is suspect and at times subject to arrest if he or she is not engaged in at least a minimal activity—going or coming, fishing, getting a tan, reading the paper, smoking, window shopping. One must always be able to make a case for one's acts having some vague utilitarian value—"broadening the outlook," "keeping up," "making contacts," "keeping in shape," "taking the mind off work for a bit," "getting some relaxation."[9] The answer to "what are you doing?" can be "nothing" only if one is a child. An adult's answer must imply some ulterior purpose— something that will be fed back into the mindless and unremitting productivity of the larger system.

This utilitarianism also underlies traditional attitudes toward pornography and drugs. In both cases the society fosters what it condemns. And both of the condemned practices threaten the harnessing of pleasure with a kind of circuit overload.

If we define pornography as any message from any public medium that is intended to arouse sexual excitement, then most advertisements are pornographic. But when we examine the specific rules concerning what is *considered* pornographic, we discover that the real issue is one of completion: the body can be only partially nude, sexual organs cannot be shown, a sexual act cannot be completed, and so on. The reason for this is that a partial arousal can be harnessed for instrumental purposes, while too strong an impulse might distract the audience from these purposes—might lead them to forsake buying for direct experience.

This leads to a self-defeating cycle. The more successful we are in generating esoteric erotic itches that can only be scratched in the world of fantasy, and thus lend themselves well to marketing—the stronger becomes the desire to obtain pure and uncontaminated gratifications. Our senses are numbed by utility, and the past fifteen years have seen an impressive flowering of techniques and movements and exhortations designed to reverse this process and to purify experience. Now, the more attractive the idea of uncontaminated experience becomes, the closer all media must approximate it before pulling back and shunting the audience off into the marketplace. But raising the ante in this way simply aggravates the need, and the whole process can only escalate until the donkey either gets his carrot or runs amok. The "relaxation" of restrictions on sexuality in the media is not a relaxation at all, but merely another intensification of the control-release dialectic on which Western civilization is so unfortunately based.

Critics of censorship are fond of pointing out that censors are strangely tolerant of violence—that it's perfectly all right for a man to shoot, knife, strangle, beat, or kick a woman so long as he doesn't make love to her. An irate father taking his children to a "family movie" (consisting of brutal and bloody killings and the glorification of hatred) complained bitterly when they were exposed to previews that showed some bare flesh and love-making. The children must be trained into our competitive value system in which it is moral for people to hurt one another and immoral for them to give pleasure to one another.

Lenny Bruce used to point out that a naked body was permissible in the mass media as long as it was mutilated. This is true but for a very good reason: our society needs killers from time to time—it doesn't need lovers. It depends heavily on its population being angry and discontented; the renunciation of violence would endanger our society as we now know it. Failure to do so, of course, endangers *us,* but our society doesn't cater to people. The reason a mutilated body is more acceptable than a whole one is that it's only in mutilated form that the sexual impulse has been allowed to exist in America. In pure form it would dissolve our culture and consign its machinery to rust and ruin, leaving a lot of embarrassed people alone with each other.

Even the "sexual revolution" has been contaminated by male preoccupations with competition and achievement. Most American men aren't really interested in pleasurable stimulation—they want ego-boosting and tension release. It's not unfair to say that a large proportion of American men spend the brief time that they inhabit their bodies looking for the "Off" button. Whereas women are usually concerned with the quality of a sexual experience, men tend to focus on quantity. They think they're having a good sex life if they're successful in turning *women* on, even if their own sensations are minimal. As Alexander Lowen points out, many people go through life never knowing what a rich sexual experience is like, since they have nothing to compare their own with. If they get some genital pleasure and a local orgasmic release, they define their sex life as "good," even when sensation is numbed and limited. Only when they break through to new levels do they realize how pallid "good" can be. This is particularly true for men, since they are so often preoccupied with performance and getting a big response from women. Norman Mailer, for example, once wrote a long, laborious story about getting a woman to have an orgasm, while the man's pleasure was scarcely mentioned. Most men have difficulty *receiving* pleasure, unless it proceeds quickly to orgasm.

Taking drugs involves the same ambivalence. Marijuana and psychedelics, for example, offer a return to pure experience, to unencumbered sensation. People take them to encounter the world in terms of what it is rather than how it can be used ("this beach will be a valuable resort property some day"). They want to stop treating themselves as machines.

Yet the means they employ involve just that. Drugs, like pornography, are both a logical development from, and a reaction against, our culture. They're attacked for blowing the mind instead of just tickling it, but in the last analysis they merely raise the ante, and ultimately the marketplace will incorporate and exploit them.

People who take drugs regularly to alter their consciousness in some way are behaving like good American consumers. The mass media are always telling us to satisfy our emotional needs with material products—particularly through oral consumption. Our economy depends upon our willingness to turn to things rather than people for gratification—to symbols rather than our bodies. The gross national product will reach its highest point when a material object can be interpolated between every itch and its scratch.

Training for this begins even before television. Mothers are always advised that if their two-year-old masturbates, they should take her hand away and give her a toy, and most parents would prefer to have their child sucking a pacifier rather than a thumb, and clutching a blanket rather than a penis. Blankets and pacifiers, after all, can be bought in a store, thus reducing unemployment.

The drug world simply extends this process in its effort to reverse it. If the body can be used as a working machine, and a consuming machine, why not an experience machine? Drug consumers make the same assumption as all Americans—that the body is some sort of appliance. Hence they must turn on and tune in, in their unsuccessful effort to drop out. They may be enjoying the current more, but they're still plugged into the same machinery that drives other Americans on their weary and joyless round.

Enjoy, Enjoy

These examples help explain why the mass media in our society seem so omnivorous—devouring and trivializing each new bud of change almost before it can fully emerge. I'm always reminded of those science-fiction monsters that "eat" radioactivity and must constantly seek new sources of this energy. What the mass media eat is new forms of emotional expression. The more the sexual impulse is exploited instrumentally, the more "valuable" it becomes economically. The act of buying has become so sexualized in our society that packaging has become a major industry: we must even wrap a small purchase before carrying it from the store to our home. Carrying naked purchases down the street in broad daylight seems indecent to Americans. After all, if we're induced to buy something because of the erotic delights that it covertly promises, then buying becomes a sexual act. Indeed we're approaching the point where it absorbs more sexual interest than sex itself; when this happens people will be more comfortable walking in the street nude than with an unwrapped purchase. Package modesty has increased in direct proportion as body modesty has lessened.

But sexuality as a marketing resource is not inexhaustible. In the absence of real gratification, interest threatens to flag, and the search for new raw material is an increasingly desperate one. New images, new fantasies of an exciting, adventurous, and gratifying life must be activated. Even efforts to change the direction of the society are gobbled up to further titillate and excite the product-filled discontent that prevails.

Eldridge Cleaver promised that blacks would rescue America from all this by a kind of emotional transfusion. While Freud called man "a kind of prosthetic God" whose auxiliary organs had not quite grown onto him yet, Cleaver argued that man needed "an affirmation of his biology" and "a clear definition of where his body ends and the machine begins." He thought that "blacks, personifying the Body and being thereby in closer communion with their biological roots than other Americans," could provide this affirmation.[10] To some extent this has happened, although the effect has been limited by numbers and prejudice.

How much it's needed is apparent when one listens to Western employers in developing nations complaining that their workers haven't learned "rational" attitudes toward machinery. Upon probing further we discover that these "rational" attitudes consist in a) acting as if one owned the machine, and b) treating it as a person. Our Western view is apparently that animism is rational when it pertains to inanimate manmade objects but irrational and "primitive" when it pertains to living things. If non-Western workers need more libidinal involvement with machines, it seems clear that Americans could do with a lot less.

Women, who are less alienated than men are from their emotional life, from their vulnerability, and from sensual experience, can play a role somewhat like the one Cleaver assigns to blacks. But once again, there are two ways of looking at this reaffirmation of the body. Is it a new and saving force? Or is it merely more libidinal raw material in the process of being gobbled up by the ravenous monster that fuels our society? Won't it all end, like the hippie movement in the sixties, as a source of additional sensual titillation, designed to inflame Americans into further frantic buying and demented striving? Will these forces free the donkey, or just provide a more exotic carrot? Can they rescue Americans, as Isis rescued Lucius in *The Golden Ass* from their dreams and their machines?

FIVE

Divided We Sit

*. . . we have exalted some of the most distasteful
of human qualities into the position of the
highest virtues.*

KEYNES

*Sometimes the light's all shining on me
Other times I can barely see*
HUNTER, GARCIA, LESH, WEIR

In the late sixties there was a lot of talk about a counter-culture
in the United States. People used the word in a way that con-
jured up images of an alternative society emerging to overwhelm
the existing one. Around 1972 this mythical organism was pronounced dead by those who had invented it, who then moved on
to announce new social moods and movements in their continuing struggle with public apathy.

Style-setters have their uses, but when it comes to understanding social change, they're just flies on the lens of a tele-
scope. Blue jeans and tuxedos, activism and conformity, have
been pronounced "in" and "out" many times during the last fifty
years, yet none of them has stopped happening. What makes
"news" and what makes change are two completely different
things.

Two things *did* happen in the sixties, and both are continuing.
The first was a strong attitudinal shift among part of the population—the affirmation of a set of values diametrically opposed to
those of the main part of society. The second was the emergence
of some alternative institutions—communes and work collec-

tives, organizations among the oppressed, and so on—trying to provide services and fulfill needs without buying into the competitive assumptions of the society as a whole.

The alternative institutions were small and fragile. They still are. In the sixties their number and strength were exaggerated because they were newsworthy. Now they're ignored and presumed to be defunct because they're *not* newsworthy. The only kind of growth that attracts the media is a "rage"—a fad that sweeps the country and for that very reason has to be short-lived. Anything long-range produces boredom. This is why the media must inevitably miss any significant social change: all important social change is too slow to arouse their interest.

The same holds for the attitude change. In the sixties the assertion of countercultural attitudes was very new and publicly stimulating. Today those attitudes are no longer new and hence of no media interest. Surveys of college students seem to agree that attitudes haven't changed since the sixties—the peace and quiet on the campuses has more to do with cynicism about the usefulness of activism and protest. Since the number of radical and turned-on students in the sixties was greatly exaggerated, what researchers refer to as "no change" probably represents an increase in countercultural values. Even protests and demonstrations have by no means disappeared, but they tend to be smaller in scale and get little attention in the news.

In short, there are more people in their twenties holding countercultural values than there were ten years ago, and they're living lives more influenced by those values. The fanfare, explosions, and confrontations, however, are gone. Had the activists of the sixties either won the day or been eradicated, then it would be reasonable to speak of "the counterculture" as deceased. But life isn't that simple. The issues that people got so excited over in the sixties are still unresolved. The conflicts of that period go to the heart of American culture. Some of what "the counterculture" was selling has been bought. Some attitudes and lifestyle changes are no longer controversial because they've become absorbed, in diluted form, into the mainstream of the culture. Some of the very people who pronounced "the counterculture" dead were visibly altered by it.

The more important parts of "the counterculture"—its radical attack on competition and the concentration of wealth and power have subsided into attempts to build alternative institutions outside the society's mainstream. These efforts are being ignored, since they are small and nonthreatening. The mass protests and demonstrations which helped give people hope—a feeling that they weren't alone but part of a great movement—are gone. People are struggling along as best they can, feeling like a tiny minority. No one knows how many will emerge from the woods when the next big issue arises.

For the sixties were just the first battle in a struggle that will last for many decades. It wasn't entirely futile: two exceptionally arrogant presidents were toppled and a war ended—a war in which national prestige was heavily involved. Democracy was much revitalized: the House UnAmerican Activities Committee—sacrosanct for almost two decades—died, and other previously invulnerable threats to democratic process—the military budget, the CIA, the FBI—came under attack. Seldom in the past century has the public been so testy in the face of official arrogance, and "the counterculture" can take much of the credit for this.

Yet the structure of our society—its massive inequalities, overwhelming corporate power, and the chronic exploitation of the consumer by professions, industry, and government—has not changed. It will take more than enthusiasm to bring that about.

What the turmoil of the sixties did was to define the issues of the struggle. It became clear that two opposing cultural systems were coming into conflict: if "the counterculture" was overbilled as a unified social force, it nevertheless represented a new cultural idea (new in our society, at least). In chaotic, disorganized form it offered a new culture to Americans. It was premature, sudden, a mood only, and the mood passed. But the struggle between the two cultures will continue for generations.

I talked in the last chapter about the mass media devouring novelty, and radicals often talk about the establishment "co-opting" them. Some people feel that the old culture simply gobbled up "the counterculture" and went about its business after a

94   CHAPTER 5

belch or two. But social change is more complicated than these simple-minded ideas of struggling adversaries. Social change doesn't happen because David slays Goliath. David and Goliath are inside everyone, and inside every one of our institutions as well. No revolution is ever over: it's never won and it's never lost. Capitalism isn't dead in the Soviet Union and collectivism still breathes in the United States. If David kills Goliath in the government, the fight still goes on in the market, or in the schools, or in medicine, or in the family; and if killed in all those places, it still goes on inside every individual's head. The "counterculture" itself was rife with individualism and riddled with capitalism, as its opponents were so fond of pointing out.

The point is that if the new culture was devoured by the old, the battle will continue in the old culture's belly. At some point a devourer always overreaches himself, like the witch or giant in folk tales who tries to drink up the sea and bursts, or like the vacuum monster in *Yellow Submarine* who ultimately devours himself and disappears.

The Two Cultures

A culture is largely a symbolic thing—a set of values, ways of behaving, ways of looking at the world, ideas, beliefs, designs. It is not a collection of people. (The word "counterculture," however, was often used to refer to collections of people—students, New Left radicals, communards, flower people, and so on, which is why I put it in quotation marks.) When I talk about two cultures in the United States, then, I'm not referring to warring groups of people, but rather to colliding idea systems. Each system is more or less consistent within itself. Each is based on a set of assumptions and has an internal logic that hangs together as long as those assumptions hold. Hence both are valid, although they are diametrically opposite and cannot be combined.

The basic assumption of the old culture is that human gratification is in short supply; the new culture assumes that it is plentiful. Like all basic assumptions, they're both self-validating—that is, whichever one you believe and act on tends to be true.

Based on these assumptions, the old culture tends to choose property rights over personal rights, technological requirements over human needs, competition over cooperation, violence over sexuality, concentration over distribution, producers over consumers, means over ends, secrecy over openness, social forms over personal expression, striving over gratification, Oedipal love over communal love, and so on. The new culture tends to reverse all these priorities.

Now it's important to realize that these differences can't be resolved by some sort of compromise. Every cultural system is a dynamic whole, and change must affect the motivational roots of a society or it isn't change at all. If you introduce some isolated new element into such a system, it will either be redefined and absorbed if the culture is strong, or will have a disorganizing effect if the culture is weak. As Margaret Mead points out, to introduce cloth garments into a grass-clad population without simultaneously introducing closets, soap, sewing, and furniture, merely transforms a neat and attractive tribe into a dirty and slovenly one. Cloth is part of a complex cultural pattern that includes storing, cleaning, mending, and protecting—just as the automobile is part of a system that includes fueling, maintenance, and repair. You can't just graft lungs onto a shark and expect it to survive out of water.

Imagine, for example, that we're cooperation freaks, trying to remove the competitive element from foot races. We decide, first of all, that we'll award no prize to the winner, or else prizes to everyone. This, we discover, fails to reduce competitiveness. Spectators and participants alike are still preoccupied with who won and how fast he or she ran relative to someone else, now or in the past. We then decide to eliminate even *announcing* the winner. To our dismay we discover that our efforts have created some new customs: the runners have taken to wearing more conspicuous clothing—bright-colored trunks and shirts, names emblazoned in iridescent letters—and underground printed programs have appeared with names, physical descriptions, and other information to help spectators identify them. In despair we decide to have the runners run one at a time and we keep no time records. But now we find that the sale of stopwatches has become a booming business, that the underground printed programs have expanded to include statistics on past time-records of runners, and that private "timing services," like the rating services of the television industry, have grown up to provide im-

mediate time results for spectators willing to pay a nominal sum
(thus does artificial deprivation encourage free enterprise).

At this point we're obliged to eliminate the start and finish
lines—an innovation that arouses angry protest from both runners and spectators, who have made only mild grumblings over
our previous efforts. "What kind of a race can it be if people
begin and end wherever they like? Who will be interested in it?"
To mollify their complaints and combat dwindling attendance we
reintroduce the practice of having everyone run together. Soon
we find that the runners are all starting at once again (although
we disallow beginning at the same place) and that all races are
being run on the circular track. The races get longer and longer,
and the printed programs now record statistics on how many laps
were run by a given runner in a given race. All races have now
beome endurance contests and one goes to them equipped with
a picnic basket. The newer fields, in fact, don't have bleachers,
but only tables at which drinks are served, and scattered TV
monitors at which the curious look from time to time and report
to their tables the latest news on surviving runners.

Time passes, and we're increasingly subjected to newspaper
attacks concerning the corrupt state into which our efforts have
fallen. With great trepidation we inaugurate a cultural revolution
and make further drastic changes in the rules. Runners begin and
end at a signal, but there is no track, just an open field. A runner
must change direction every thirty seconds, and anyone who
runs parallel with another runner for more than fifteen seconds
is disqualified. At first attendance falls off badly, but after a time
spectators become interested in how many runners can survive
a thirty-minute race without being eliminated for a breach of the
rules. Soon some running teams become so skilled at not running
parallel that none of them are ever disqualified. They begin to
develop intricate patterns of synchronizing their direction
changes. The more gifted teams become virtuosi at moving parallel until the last split second and then diverging. The motions
of the runners become more and more elegant, and a vast out-
pouring of books and articles descends from and upon the university (ever a dirty bird) to establish definitive distinctions between the race and the dance.

The first half of this parable is a portrait of what most liberal reform amounts to: opportunities for the established system to flex its muscles and exercise its self-maintaining capabilities. Poverty programs put very little money into the hands of the poor because middle-class hands are so much more gifted at grasping money—they know better where it is, how to apply for it, how to divert it, how to concentrate it. That's what being middle-class means, just as a race means competition.

No matter how much we try to change things they somehow end as just a more intricate version of what existed before. A heavily graduated income tax ends by making the rich richer and the poor poorer. Most of the money raised in a charity drive goes to middle-class fundraisers, middle-class administrators, and middle-class professionals. When a government program is created to provide psychiatric services for the poor, a clinic that had charged welfare recipients five dollars an hour begins to charge thirty dollars an hour, since the government is paying for it. Hospitals sometimes support themselves by filling empty beds with state-funded welfare clients, on whom unnecessary operations are performed by residents and interns who need the practice. People in movie theaters see heart-rending film appeals to help poor children crippled with various diseases, but the money collected goes to subsidize middle-class professionals doing obscure academic research to advance their own personal careers. If you empty a sack of money into a crowd, the greediest will get the most, and no matter how many nipples a government puts out for the poor, it always seems to find a middle-class mouth fastened around every one.

But there's a limit to how much change a system can absorb, and the second half of our parable suggests that if we persist in our efforts and attack the system at its motivational roots, we may indeed be successful. In any case, there's no such thing as a compromise: we're either strong enough to lever the train onto a new track or it stays on the old one.

Core Beliefs

Thus it's important to know the motivational foundations of the old and the new cultures, for a prolonged head-on collision would nullify both of them, like bright pigments combining into gray. The transition must be as deft as possible if we're to minimize the destructive chaos that accompanies major cultural transformations. There must be room for them to exist side by side while the energy is gradually siphoned from one to the other. The complexity and richness of our society lends itself to this, and it's happening today in a small but significant way.

The core of the old culture is scarcity. Everything in it rests on the assumption that the world does not contain the wherewithal to satisfy the needs of its human inhabitants. From this it follows that people must compete with one another for these scarce resources—lie, swindle, steal, and kill, if necessary. These basic assumptions, however, create the danger of a "war of all against all" and must be qualified by moral commandments that will mute and restrain the intensity of the struggle. Those who can take the largest share of the scarce resources are said to be successful, and if they can do it without violating the moral commandments, they are said to have character and moral fibre.

The key flaw in the old culture is that so much of its scarcity is spurious—that it creates most of the needs it fails to satisfy and often deliberately frustrates others. We suffer from an energy shortage, for example, because we've learned to "need" cars, the presence of which frustrates our more basic needs for community, stability, closeness, warmth, and physical contact. Even world hunger is to a considerable extent man-made, by maldistribution of food, overpopulation, and dependence on agriculture, which in turn comes from human anxiety to create food surpluses. Malnutrition and famine are less common in hunter-gatherer societies, for example, than in agricultural ones. During severe droughts agricultural peoples sometimes have to turn to their "poor" neighbors in order to survive. Since they have few wants, many "poor" hunter-gatherers feel rich indeed.[1]

Most scarcity in our own society exists for the purpose of maintaining the system that depends upon it. Americans are often in the position of having killed someone to avoid sharing a meal which turns out to be too large to eat alone.

The new culture is based on the assumption that important human needs are easily satisfied and that the resources for doing so are plentiful. Competition is unnecessary, and the only danger to humans is human aggression. There is no reason outside of human perversity for peace not to reign and for life not to be spent in joy and the cultivation of beauty. Those who can do this in the face of the old culture's ubiquity are considered "beautiful."

The flaw in the new culture is the fact that the old culture has succeeded in making real the scarcity it believes in—that a certain amount of work will be needed to release all this bounty from the restraints under which it is now placed. A man may learn that his nearsightedness is "all psychological" and still need his glasses to see. And should our transportation system break down, New York City will starve to death, despite the abundance of food around the nation.

It's important to recognize the internal logic of the old culture, however faulty its premise. If you assume scarcity, then the knowledge that others want what you want leads with some logic to preparations for defense, and ultimately (since the best defense is offense) for attack. The same assumption leads people to place a high value on the ability to postpone gratification (since there's not enough to go around). The expression of feelings is a luxury, since this might alert the scarce gratifications to the fact that a hunter is near.

Stimulation

Self-restraint and coldness (which create still further scarcity) lead to another norm, that of "good taste." One can best understand a norm like this by seeing what's common to those acts that violate it, and on this basis the meaning of "good taste" is very clear—"good taste" means tasteless in the literal sense. Any act or product which contains too much stimulus value is considered to be "in bad taste" by the old culture. Since gratification is a scarce commodity, arousal is dangerous. Clothes must be drab and inconspicuous, colors of low intensity, smells nonexistent. Sounds should be quiet, words should lack emotion. Four-letter words are in bad taste because (to older people, at least) they have high stimulus value. Satire is in bad taste if it arouses political passions or creates images that are too vivid or exciting. Brilliant, intense, vibrant colors are called "loud," and the preferred

colors for old-culture homes are dull and listless. Stimulation in
any form leaves old-culture Americans with a "bad taste" in their
mouths. This taste is the taste of desire—a reminder that life in
the here-and-now contains many pleasures to distract them from
the carrot dangling beyond their reach. Too much stimulation
makes the carrot hard to see. Good taste is a taste for carrots.

In the sixties the "counterculture" began, with much fanfare,
to develop a style in direct opposition to all this—a style that
now dominates some segments of our society, coexisting peacefully with the older one. For if you assume that gratification is
easy and resources plentiful, stimulation is no longer to be
feared. Psychedelic colors, amplified sound, erotic books and
films, bright and textured clothing, spicy food, "intense" (i.e.,
Anglo-Saxon) words, angry and irreverent satire—all run
counter to the old pattern of understimulation. Longer hair and
beards provided a more "tactile" appearance than the bland, geometric lines of the fifties. Even Edward Hall's accusation that
America is a land of "olfactory blandness" (a statement any traveler will confirm) had to be qualified a little. Smell is the sense
that the old culture feels most violently about muting, however,
and when America is filled with intense color, music, and orna-
ment, deodorants will be the old culture's last bastion.

The old culture, then, turned the volume down on emotional
experience in order to concentrate on dreams of glory. The new
culture turned it up again, to drown out those dreams. New-
culture people, in fact, often show signs of *under*sensitivity to
stimuli. They seem to be more certain that desire can be gratified than that it can be aroused—a response that probably owes
much to Spockian child-rearing. In earlier times a mother re-
sponded to her child's needs when they were expressed powerfully enough to distract her from other cares and activities.
Spockian mothers, however, often tried to anticipate the child's
needs: before arousal had proceeded very far they hovered about
offering several possible satisfactions. Since we tend to use early
parental responses as models for the way we treat our own impulses in adulthood, some people find themselves moving toward gratification before the need itself is clear or compelling.
Like their mothers, they aren't altogether clear which need

they're feeling. To make matters worse they're caught in the dilemma that spontaneity evaporates the moment it becomes an
ideology: it's a paradox of modern life that those who oppose libidinal freedom may be more capable of achieving it.

Inequality

Another logical consequence of scarcity is structured inequality.
If there isn't enough to go around, then those who have more
will find ways to prolong their advantage, and even try to legiti-
mize it. The law itself, although equalitarian in theory, is in large
part a social device for maintaining structured inequality—keeping the rich rich and the poor poor. One of the major thrusts of
the new culture, on the other hand, is equality: since the good
things of life are plentiful, everyone should share them—rich
and poor, black and white, female and male.

In the old culture, means habitually become ends, and ends
means. Instead of people working to obtain goods in order to be
happy, for example, we find that people should be made happy
in order to work better in order to obtain more goods. Inequality,
originally a result of scarcity, is now a means of *creating* it. And
for the old culture, the manufacture of scarcity is paramount, as
hostile comments toward new-culture customs ("people won't
want to work if they can get things for nothing") reveal. Scarcity,
the supposedly unfortunate but unavoidable foundation for the
whole system, has now become its most treasured and sacred
resource, and in order to maintain this resource in the midst of
plenty, it has been necessary to make invidiousness the foremost
criterion of worth. Old-culture Americans are peculiarly drawn
to anything that seems to be the exclusive possession of some
group or other, and they find it difficult to enjoy anything unless
they can be sure that there are people to whom this pleasure is
denied. For those in power, life itself derives value invidiously.
Numbed by their busy lives, many officials gain reassurance of
their vitality from their proximity to blowing up the world.

Invidiousness also provides a *raison d'être* for the advertising
industry, whose primary function is to manufacture illusions of
scarcity. In a society engorged to the point of strangulation with
useless and joyless products, advertisements show people calam-

itously running out of their food or beer, avidly hoarding potato chips, stealing each other's cigarettes, guiltily borrowing each other's deodorants, and so on. In a land of plenty there should he little to fight over, but in the world of advertising men and women will fight before changing their brand, in a kind of parody of the Vietnam war.

The fact that property takes precedence over human life in the old culture also follows logically from scarcity assumptions. Since possessions are "scarce" relative to people (in the sense that there are always some people who don't own a given form of property), they come to have more value than people. This is especially true of people with few possessions, who come to be considered so worthless as to be subhuman and eligible for extermination. Having many possessions, on the other hand, entitles the owner to a status somewhat more than human. But as a society becomes more affluent, these priorities begin to change—human life increases in value and property decreases. Yet it's still permissible to kill someone who's stealing your property under certain conditions—especially if that person is without property himself. Thus although the death of a wealthy kleptomaniac, killed while stealing, would probably be thought worthy of a murder trial, that of a poor black looter would not. Noise control provides another example: Police are called to prevent distraction by the joyous noises of laughter and song, but not to stop the harsh and abrasive roar of power saws, air hammers, power mowers, snow blowers, and other baneful machines.

Old Loves New, New Loves Old

It would be easy to show how various aspects of the new culture derive from the premise that life's satisfactions exist in abundance and sufficiency for all, but I would like to look instead at the relationship the new culture bears to the old—the continuities and discontinuities it offers.

To begin with, the new culture takes the position that instead of throwing away your body in order to accumulate possessions, you should throw away the possessions and enjoy your body. This is based on the idea that possessions actually create scarcity. The

more emotion you invest in them, the more chances for real gratification are lost—the more committed to them you get, the
more deprived you feel, like a thirsty man drinking salt water.
To accumulate possessions is to deliver pieces of yourself to dead
things. Possessions can absorb emotion, but unlike people, they
feed nothing back. Americans have combined love of possessions
with the disruption and trivialization of most personal relationships. An alcoholic becomes malnourished because drinking obliterates his hunger; Americans become alienated because
amassing possessions obliterates their loneliness. This is why
manufacturing in the United States has always seemed to be on
an endless upward spiral: every time we buy something we
deepen our emotional deprivation, and hence our need to buy
something. This is good for business, of course, but those who
profit from it are just as trapped as everyone else. The new culture seeks to substitute an adequate emotional diet for this crip-
pling addiction.

Yet the new culture is a product of the old, not just a rejection
of it. It picks up themes that are dormant or secondary in the old
culture and magnifies them. It's full of nostalgia—nostalgia for
the Old West, Amerindian culture, the wilderness, the simple
life, the utopian community—all venerable American traditions.
But for the old culture they represent a minor aspect of the culture, appropriate for recreational occasions or fantasy—a kind of
pastoral relief from everyday striving—whereas for the new culture they're dominant themes. The new culture's passion for
memorabilia, paradoxically, causes uneasiness in old-culture adherents, whose striving leads them to sever themselves from the
past. Yet for the most part the new culture is simply making the
old culture's secondary themes primary, rather than just discard-
ing the old culture's primary theme. Even the notion of "dropping out" is an important American tradition—neither the
United States itself nor its populous suburbs would exist were
this not so.

Americans have always been deeply ambivalent about social
involvement. On the one hand they're suspicious of it, and share
romantic dreams of running off to live in the woods. On the other
hand they're much given to acting out grandiose fantasies of tak-

ing society by storm through the achievement of wealth, power, or fame. This ambivalence has led to many strange institutions— the suburb and the automobile being the most obvious. But note that both fantasies take the viewpoint of an outsider. Americans have a profound tendency to feel like outsiders—they wonder where the action is and wander about in search of it (this puts a great burden on celebrities, who are supposed to know, but in fact feel just as doubtful as everyone else). Americans have created a society in which they are automatic nobodies, since no one has any stable place or enduring connection. The village idiot of earlier times was less a "nobody" in this sense than the mobile junior executive or academic. An American has to "make a place for himself" because he does not have one.

Since our society rests on scarcity assumptions, involvement in it has always meant competition, while oddly enough, the idea of a cooperative, communal life has always been associated with bucolic withdrawal. So consistently have cooperative communities established themselves in the wilderness, we can only infer that society as we know it makes cooperative life impossible.

It's important to remember that the New England colonies grew out of utopian communes, so that the drop-out tradition is not only old but extremely important to our history. Like so many of the more successful nineteenth century communities (Oneida and Amana, for example), the puritans became corrupted by economic success until the communal aspect eroded away—another example of a system being destroyed by what it tries to ignore. The new culture is thus a kind of reform movement, trying to revive a decayed tradition once important to our civilization.

It Might Come in Handy

These continuities between the new culture and the American past reflect a characteristic of all successful societies—perhaps all living organisms: they all have some method of keeping alive ideas that don't jibe with the dominant emphasis of the status quo. It's a kind of hedge against future changes. These latent alternatives usually survive in some encapsulated and imprisoned form ("break glass in case of fire"), such as myths, festivals, or deviant subgroups. Fanatics are always trying to expunge these contradictions, but when they succeed it's usually fatal to the society. As Lewis Mumford once pointed out, it's the disorder in a system that makes it viable, considering the contradictory needs that all societies must satisfy.[2] Such inconsistencies are priceless treasures and must be carefully guarded. For a new cultural pattern doesn't emerge out of nothing—the seed must already be there, like the magic tricks of wizards and witches in folklore, who can make an ocean out of a drop of water, a palace out of a stone, a forest out of a blade of grass, but nothing out of nothing. Many peoples keep alive a tradition about a golden age in which a totally different social structure existed. Jesters kept playfulness alive amid the formality of royal courts. The transvestite roles in many warrior cultures were a living reminder that *macho* rigidity wasn't the only conceivable kind of behavior for a male; conversely, the warrior ethic is maintained in peaceful societies by military cadres. Today, anthropology helps fulfill these needs, showing us many different ways of coping that contrast sharply with our own.

These alternative social practices are like a box of seldom-used tools, or a trunk of old costumes awaiting the proper period-play. Suddenly the environment changes, the eccentric becomes a prophet, the clown a dancing-master, the doll an idol, the idol a doll. The elements haven't changed, only the arrangement and the emphases have changed. Every revolution is in part a revival.

Sometimes the society's ambivalence is so strong that the secondary pattern is preserved in a form almost as elaborate as the dominant one. Our society, for example, is one of the most mobile ever known; yet, unlike other nomadic cultures, it makes little concession to this mobility in its handling of possessions. Our homes are furnished as if we intended to spend the rest of our lives in them, instead of moving every few years. Perhaps it's just a yearning for stability expressed in a failure to adapt, but should Americans ever settle down, they won't have to do much about readjusting their household furnishing habits. Urban Moroccans, on the other hand, still use portable, stored furnishings that they assemble for each occasion, just as if they were living in tents, although in fact they've been settled in permanent dwellings for five hundred years.

Neoteny

This tendency of human societies to keep alternative patterns alive has biological analogues. One of these is *neoteny*—an evolutionary process in which what were previously fetal or juvenile characteristics are retained in an adult animal in response to altered environmental circumstances. (Many human features, for example, resemble the fetal traits of apes.) I didn't choose this example at random, for much of the new culture is "neotenous" in a cultural sense: behavior, values, and lifestyles formerly seen as appropriate only to childhood are being retained into adulthood as a counterforce to the old culture.

I pointed out earlier, for example, that children are taught a set of values in earliest childhood—cooperation, sharing, equalitarianism—which they begin to unlearn as they enter school, where competition, invidiousness, status-seeking, and authoritarianism prevail. By the time they enter adult life, children are expected to have largely abandoned the values with which their social lives began. But for affluent, protected, middleclass children, this process is slowed down while intellectual development is speeded up, so that childhood values can become integrated into a conscious, adult value system. The same is true of other characteristics of childhood: spontaneity, hedonism, candor, playfulness, use of the senses for pleasure rather than utility, and so on. The child-oriented middle-class family allows the child to preserve some of these qualities longer than is possible under more austere conditions, and her intellectual precocity makes it possible for her to integrate them into an ideological system with which she can confront the corrosive, life-abusing tendencies of the old culture.

When these neotenous characteristics become visible to old-culture adherents, the effect is painfully disturbing, for they vibrate with feelings and attitudes that are very old and very deep, although long and harshly stifled. They have learned to reject these attitudes, but since the learning predated their intellectual maturity, they have no coherent ideological framework within which this rejection can be consciously thought out. They are deeply attracted and acutely revolted at the same time, and they can neither resist their fascination nor control their antipathy. Usually such rote-learned abhorrence has to discharge itself in persecution before people can tune in to their more positive responses. This was true in the case of early Christianity in Rome. The persecution makes it easier for the values of the oppressed group to be expropriated by the majority and released into the mainstream of the culture. The "disappearance" of the counterculture is an example of the same process.

SPLITS IN THE NEW CULTURE

1. Inner and Outer

Up to this point we have rather awkwardly discussed the new culture as if it were integrated and monolithic, which it certainly is not. There are many contradictory streams feeding the new culture, and some of these deserve particular attention since they provide the raw material for future conflict.

The most important split in the new culture is that which divides outward, political change from internal, psychological transformation. The first requires confrontation, revolutionary action, and radical commitment to changing the structure of modern industrial society. The second involves a renunciation of that society in favor of the cultivation of inner experience, psychic balance, or enlightenment. Heightening sensory receptivity, committing oneself to the here-and-now, and attuning oneself to the physical environment are also sought, since in the old culture immediate experience is overlooked or grayed out by the preoccupation with utility and mastery.

Since political activism is task-oriented, it partakes of certain old-culture traits: postponement of gratification, preoccupation with power, and so on. To seek political change is to live in the future, to be absorbed in means, in achievement. Psychological change can be sought in the same way, of course—you can make anything into a task—but to do so is to treat yourself as an object to be manipulated. No significant psychological change can occur if the change is sought as a means to some other end. The change must be an end in itself. True psychological change then, whether we call it "sensory awakening," "enlightenment," "mental health," or just "getting your shit together," is a "salvation now" approach. Thus it's more radical, since it's uncontaminated with old-culture values. It's also less realistic, since it ignores the fact that the status quo provides a totally antagonistic milieu in which the psychological changes will be undermined. The "flower children" of the sixties, for example, tried to survive in a state of highly vulnerable parasitic dependence, and were severely victimized by the old culture. They had no social base of their own and evaporated at the first sign of trouble. Yet, on the other hand, an attempt at political change based on old-culture premises is lost before it's begun, for even if its authors are victorious, they will have been corrupted by the process of winning.

The dilemma is a real one wherever radical change is sought. For every society tries to exercise rigid control over the mechanisms by which it can be altered—defining some as legitimate and others as criminal or disloyal. When we look at the characteristics of these mechanisms, however, we find that the "legitimate" ones require a sustained commitment to the core assumptions of the culture. In other words, if a person follows the "legitimate" pathway, there's a very good chance that her initial radical intent will be eroded in the process. If she feels that some fundamental change in the system is required, then she has a choice between following a path that subverts her goal or one that leads to some form of victimization.

This isn't a Machiavellian invention of American capitalism, but simply the way all viable societies protect themselves from instability. When the society is no longer viable, however, this technique must be exposed for the swindle that it is; otherwise the needed radical changes will be rendered ineffectual.

The key to the technique is the powerful human reluctance to admit that an achieved goal was not worth the unpleasant experience required to achieve it.[3] This is the principle on which initiation rituals are based: "If I had to suffer so much pain and humiliation to get into this club, it must be a wonderful group." The evidence of thousands of years is that the mechanism works

extremely well. Most professionals are reluctant to admit that
their graduate training was boring, trivial, nitpicking, or irrelevant, as so much of it is, and up to some point war leaders can
count on high casualties to increase popular commitment to military adventures

So when a political leader says to a radical, "Why don't you
run for political office and try to change the system from within,"
or the teacher says to the student, "Wait until you have your
Ph.D. and then you can criticize our program," they're dealing a
stacked deck, in that the protester, if she accepts the condition,
will in most cases be converted to her opponent's point of view
rather than admit to having worked, struggled, and suffered
pointlessly.

The dilemma of the radical, then, is that she is likely to be
corrupted if she fights the *status quo* on its own terms, but is not
permitted to fight it in any other way. And even if she succeeds
in solving this dilemma, after a lifetime spent altering the power
structure, won't she become old-culture utilitarian, invidious,
scarcity-oriented, future-centered, and so on? Having made the
world safe for the enlightened, can she afford to relinquish it to
them?

There is no way to resolve this dilemma and, indeed, it's better left unresolved. Political change requires discipline and unity
of purpose, which lead to all kinds of abuses when the goal is
won. Discipline and unity become ends in themselves (the usual
old-culture pattern) and any victory becomes an empty one. It's
important to have someone living out the goals of any political
change even before it's achieved, so that the new political condition can be compared to a living reality, however small and
insignificant, instead of just to some visionary fantasy.

The new culture is amorphous, and since it no longer has any
significant spokespersons to give it a false air of purpose and
unity, almost anything we can say about it is arguable. Millions
of Americans are trying to break old-culture habits, but for every
one they break a dozen are reinforced. The new culture is very
divided about violence, for example: in the sixties nonviolence
seemed to be a central tenet, to the point where assertiveness

and anger were in danger of being crossed off the list of allowable human traits; but the violence of the establishment finally converted many dedicated pacifists to terrorism.

2. Technology

Technology is another point of division. The new culture tends to be suspicious of technology as a whole, but everyone has his favorite exception when it gets down to cases. Any large-scale assault on technology would produce a host of defenders even from within the attacking armies—each rushing to place some prized possession on the endangered appliance list. Most nature lovers, who dream of living peacefully in the wilderness after all the nasty bulldozers and jets and factories have been magically whisked away, fail to notice that their fantasy includes a stereo or a blender or a Ferrari. Americans are hopelessly enmeshed in equipment and won't escape it in the lifetime of anyone now living. Some people take almost as much equipment camping as they have in their homes; it's just more expensive and lighter weight.

Other new-culture types are more sophisticated and attack only mechanical technology, exempting all that's electronic from their ire. Even though it disrupts the environment just as badly, they argue, it doesn't create bad habits of thought: being circuitry, it reminds people of their interconnectedness and of the importance of feedback. The electronic buffs like to demonstrate how video equipment, for example, can connect people who wouldn't otherwise be connected. So far I'm unmoved by these demonstrations, which always remind me of two kids rigging up toy telephones so they can talk to each other across the room without shouting. They also talk of the marvels of videotape, where one can see oneself and alter one's behavior without the embarrassment of getting feedback directly from other people. But I'm a little suspicious of solipsistic learning. The more you can control your environment, the less you can learn from it. A videotape, like an audiotape, is just a better mirror, and insofar as mirrors are bad for us, videotape is worse. Anything that offers to take risk out of the environment is a bad bargain, since it always costs too much and puts as much risk in as it takes out.

3. *Individualism*

But the most important ideological confusion in the new culture has to do with individualism and collectivism. On this question the new culture talks out of both sides of its mouth, one moment pitting ideals of cooperation and community against old-culture competitiveness, the next moment espousing the old culture in its most extreme form with exhortations to "do your own thing." New-culture enterprises often collapse because of a dogmatic unwillingness to subordinate the whim of the individual to the needs of the group. This problem is rarely faced honestly by new-culture people, who seem unaware of the conservatism involved in their attachment to individualistic principles.

This is one reason why the new culture cannot be left entirely in the hands of the young and still prevail. Despite all attempts to reverse direction, our society is still moving away from the instinctive sense of community that villagers had in the past— still moving with dizzying speed toward greater anonymity, impersonality, and disconnectedness. By and large, the younger a person is, the less likely he or she is to have any instinctive communal responses left. Once gone, they're very difficult to re-create. Older people, on the other hand, still retain some vestigial ability to care what happens to a group or community, and this is a valuable resource for those seeking a more communal society.

The loss of communal skills in the young isn't all a result of mobility and transience. Spockian child-rearing, in the anxious form in which it was frequently carried out, often encouraged narcissism and insensitivity in its products, and this was profoundly reinforced by television. For when people created their own amusements, they were very much aware that a good time was something that happened when people *did* something—that your enjoyment was a product of the energy you invested—that it combined with other people's energy and accumulated and produced something new. In other words, when you create your own fun, you realize that you only get pleasure when you commit your whole being to a situation. Television teaches a different viewpoint. Those brought up on it often show an irritating pas-

sivity in communal situations. They want the group to amuse
them whether they put anything into it or not. Otherwise they
continually threaten to withdraw. If the group doesn't please
them, they switch the channel instead of redirecting the group
with their own energy, of which they often have very little.

Because of this weakness, young people often find it difficult
in their personal lives to avoid drifting into the same capitalistic
behavior they scorn when they see it on a large scale. This is the
weakest point in the new culture, for any social change feeds on
youth, and young people are even more unsteady on this issue
than their elders. Yet if individualism in the United States isn't
sharply diminished, everything in the new culture will be per-
verted and caricatured into simply another bizarre old-culture
product. There must be continuities between the old and the
new, but not about something so basic to the whole conflict.
Nothing will change in America until individualism is assigned a
subordinate place in the American value system—for individualism lies at the core of the old culture, and individualism is not
a viable foundation for any society in a nuclear age.

SIX

Spare Change

*I promise you that in the joy and laughter of
the festival nobody will . . . dare to put a sinister
interpretation on your sudden return to human
shape.*

APULEIUS

A doctor is given the power to treat, as well as diagnose, but this
power is withheld from the social critic. Social workers treat and
classify the *victims* of social malfunction but not the malfunctions
themselves. Access to a social system is usually rather sharply
curtailed by its custodians, who, having devoted their lives to
the acquisition of power, are understandably reluctant to relinquish it to persons not having made a comparable sacrifice. This
is probably very fortunate for all of us, since the record of intellectuals and academics in positions of power particularly when it
comes to preserving democratic freedoms—is no better than
that of the military.

What, then, can a book of this kind say about putting to use
whatever insights have been gained? Talk is cheap, and perhaps
the wisest course for a social critic at such a moment is to be
quiet, and let those who are gifted at social action make whatever
use of the analysis they can. Still, to suggest that a society is in a
disastrous state without offering any guides to action implies a
detachment so extreme as to disqualify the analysis.

Fortunately, there's no need to discuss ways of initiating
change, since change is already in motion. At the same time,
however, the pathology of the old culture is accelerating, so that
the dangers it produces grow as quickly as the possibility of res-

cue. The two cultures pull in opposite directions; our task is to optimize the transition from one pattern of cultural dominance to the other. To do this we must first explore some of the ambiguities and paradoxes of social change.

CHANGE AND ADJUSTMENT

Radicals look on gradualism with justifiable contempt, since it usually proves not to be change at all but just an exercise in conservative ingenuity. Furthermore, there's no place for gradualism in a life-or-death situation—you don't walk sedately out of the way when you're about to be run over by a truck. This is the crux of all arguments about change: if there's no crisis, then the impatience and aggressiveness of radicals is inappropriate. But if there *is* a crisis, they're showing great restraint as it is. To my mind the crisis is self-evident; there are very few institutions in our society that are functioning in a way to inspire confidence.

C. Wright Mills coined the term "crackpot realism" for the kind of short-run thinking that cannot reconsider an existing policy, no matter how disastrous. A crackpot realist is an administrator who throws away a million dollars because "you can't just junk a project we've put a hundred thousand dollars into." Crackpot realists cite "practical politics" to defend our support of tottering dictatorial regimes that have collapsed one after the other (indeed, our policy of trying to outbid the Communist world for white elephants has made our defeats look, in retrospect, like clever stratagems).

Crackpot realism also prevents us from guarding ourselves against the mortal hazards we create at home. When a danger, such as air pollution or destruction of the ozone layer, is first pointed out, nothing is done for a long time because there "isn't enough information to warrant any action." This is true even when the alleged threat might extinguish life on the planet. When "enough information" is finally obtained, as in the case of DDT and other pesticides, the poisons are gradually phased out, so as not to inconvenience those who profit from their manufacture. If we discovered arsenic in our flour bin, would we construct a "timetable" for phasing out the flour?

Emphysema victims have been murdered by smog for decades while industrial polluters have been indulged and even subsidized by federal, state, and local governments. And a mild bill, that quite reasonably requires the chemical industry to test toxic substances for potential hazards before inflicting them on an unsuspecting public, is considered controversial, and not only has been unable for four years to get through Congress, but is expected, as of this writing, to be vetoed by the President if it does.

Finally, crackpot realism argues that we must move slowly, if at all, in handling urban problems, despite the fact that urban conditions annually produce thousands of stunted minds, burnt-out cases, and "criminals." The middle-class "realist's" individualism nurtures today the man who will kill his child tomorrow. But it isn't practical in America to make drastic changes, even in an emergency.

Yet there is a sense in which all change is gradual. There's an illusory element in revolutionary change—a tendency to exaggerate the importance of the revolutionary moment by ignoring the subtle and undramatic changes leading up to that moment and the reactions, corruptions, and compromises that follow it. The revolutionary moment is like a "breakthrough" in scientific discovery. It's dramatic and exciting and helps motivate the dreary process of retooling society (or scientific thought) piece by tedious piece. It may be necessary in order for any real change to occur at all—even the kinds of changes that liberal reformers seek. The only reason for stressing the hidden gradualism in revolution is that revolutionaries typically expend much of their energy attacking those very groups whose "softening-up" work makes revolution possible.

How to Lose

These attacks are often based on the notion that correct radical strategy seeks to "make things worse" in order to encourage a confrontation between the forces of reaction and the revolutionary saviors. Liberal efforts at social amelioration are thus to be avoided as dampers on revolutionary fervor. Instead, one tries to bring about a situation so repressive and disagreeable that the masses will be forced to call on the revolutionaries, waiting in the wings. This kind of fatuous policy helped bring to power Hitler, who saw to it that the revolutionaries did their waiting in concentration camps. Provoking repression is an effective technique only if the repression itself is confused and ambivalent. The result of "things getting bad enough" is usually to intimidate and demoralize most of those who want change. Revolution usually occurs not when things get bad enough but when they begin to get better—when small improvements create rising aspirations and decrease tolerance for long-existing injustices. The "make things worse" approach not only isn't strategic, it isn't even revolutionary. The emotional logic behind it might be expressed as "if things get bad enough, They will see that it is Unfair." Radical movements are always plagued with people who want to lose, want to be put under protective custody.

This is not an argument for moderation—taking an extreme position can be a winning as well as a losing stance. But when changes in the desired direction are opposed *because* they keep things from getting "bad enough," we can assume at the very least that the attitude toward change is highly ambivalent.

The make-it-worse position is based on the same assumption as the "backlash" position, which argues that "if you go too far, They will turn against you." Both view public opinion as a kind of judicial Good Parent and exaggerate the importance of transient popular sentiment. Both underestimate the significance of prolonged exposure to new ideas. There's no such thing as a situation so intolerable that human beings must necessarily rise up against it. People can bear anything, and the longer it exists, the more placidly they will bear it. The job of the radical is to show people that things can be better and to move them directly and unceasingly toward that goal. The better things get, the more aware people become that they need not tolerate the injustices and miseries that remain. By the same token, no backlash situation is a real movement backward. "Backlash" implies that people once accepted and then came to reject change, but this is never the case. It's merely that the significance of the change—the reality of it—was not yet understood. Backlash is just part of the process of learning that change means change.

A true radical tries to change behavior and institutions—the good or bad opinion of the public is of no importance. The backlash-avoider is saying, "If we go too far, people will think badly of me." This is true but irrelevant. People always say: "Yes, change is necessary, but some of the leaders (militants, radicals) go too far." This distinction is useful, for it allows the conservative to discharge his anxiety, discomfort, and resentment onto individuals while learning gradually to accept the changes those individuals are creating. Similarly, the make-things-worse advocate is saying, "If things get bad enough, even I will look good by comparison and people will think well of me and say that I am right." Better he be thought a silly eccentric or a cutthroat, and progress be made.

Change can take place only when liberal and radical pressures are both strong. Intelligent liberals have often recognized the debt they owe to radicals, whose existence permits liberals to push further than they would otherwise have dared, all the while posing as compromisers and mediators. Radicals have been somewhat less sensible of their debt to liberals, partly because of the single-minded discipline radicals are forced to maintain, plagued as they always are by liberal backsliding and timidity on the one hand, and various forms of self-destructiveness and romantic posing on the other.

Yet liberal reforms often do much to soften up an initially rigid *status quo*—creating just those rising expectations that make radical change possible. Radicals often object that liberal programs generate an illusory feeling of movement when in fact nothing is changing. Their assumption is that such an illusion slows down movement, but it's just as likely that the reverse is true. Even an illusory sense of progress is invigorating, and whets the appetite for further advances. Absolute stagnation is enervating, and creates a feeling of helplessness and impotence. Radicals overlook the educative value of liberal reform, however insignificant that reform may be in terms of institutional change.

Liberal reform and radical change are thus complementary rather than antagonistic. Together they make it possible continually to test the limits of what can be done. No one group can possibly fulfill both functions—coalitions between liberals and radicals usually weaken both groups. Yet it's important to recognize that their political goals often overlap even though their personal psychological goals are almost always in conflict.

Mind and Matter

There is an even more serious split among those seeking social change: the split between those who want to change institutions and those who want to change attitudes or motives. One group wants to redirect our striving toward social goals—to build a revolutionary new society instead of empires and fortunes—while the other wants to abolish the striving itself. One wants to remake the world so it will be tolerable for us to live in; the other wants to cure us of our need to remake the world.

Positions on this question tend to be based on whether you think motivation creates institutions or vice versa. The first task of a system is to maintain itself, and every institution must therefore reactivate continually the motivational eccentricities that gave rise to it in the first place. Still, one can't avoid a feeling of skepticism when it's proposed that institutional change alone will bring about motivational change. Closing down gambling casinos never ended gambling, and closing down banks won't stop hoarding or greed. Institutions, like technology, materialize the fantasies of past generations, and unless there's reason to believe these fantasies have changed, there is little point in trying to change the institutions, since they will simply re-emerge. On the other hand, one can no longer approach the problem psychologically once the fantasies have achieved institutional form, since they now represent a reality, one in which subsequent fantasies will be rooted.

Motivation and institutions are thus twinned, like the hedgehog and his wife in the folktale, and those who would bring about change are like the frantic hare, who, racing to best the one, finds he has been outdistanced by the other. Change can take place only when institutions have been discredited and disassembled, and the motivations that gave rise to them redirected onto new pathways to gratification. Any change that ignores either of these two contradictory approaches will be short-lived or illusory.

Radicals must learn to live with such contradictions. They can't afford the luxuries of intellectuals, who are much too fond of playing out a romantic fantasy in which they, as lonely heroes, battle bravely against a crass multitude and/or a totalitarian social structure. We are no more likely than anyone else to recognize the ways in which our own behavior creates the forces that plague us from outside; in all private myths the hero is an injured innocent.

But the impersonal machinery that threatens, benumbs, and bureaucratizes the helpless individual in books of social criticism isn't something external to the individual; it is the individual—the grotesque materialization of his turning away. Herbert Marcuse quotes René Dubos on the importance of "the longing for quiet, privacy, independence, initiative, and some open space," and suggests that capitalism not only prevents it from being gratified but also numbs the longing itself.[1] It's not clear on what basis he decides that the longing for privacy is numbed in our society—one would be hard put to find a society anywhere in which the search was more desperate, or produced a greater wealth of cultural inventions, largely self-defeating. The longing for quiet, privacy, independence, initiative, and open space is a foundation-stone of American society—of the suburb, the highway, and the entire technological monstrosity that threatens to engulf us. The longing itself is not a biological need, but a response to cultural indoctrination, crowding, and social dislocation. Those who live in stable preindustrial communities have far less privacy and far less desire for it than we do. They feel less manipulated and intruded upon because they can predict and influence their daily social encounters with greater ease. The longing for privacy is created by the drastic conditions that the longing for privacy produces.

Let's take a simple mathematical example. Assume that two people have roughly equal needs for privacy and for company. If they operate individualistically with each other they'll have their needs met half the time—coming together when both want to, and being alone when both want to—and half the time they'll be miserable and frustrated, feeling either lonely and rejected or annoyed and intruded upon. If they operate cooperatively, however, and tune into each other's rhythms, they can reduce their misery time to something close to zero. The only way to ensure privacy in a crowded world, paradoxically, is through negotiation and collaboration.

Individualism, in other words, is a narcotic, and when its virtues are touted by those in power, it's a useful divide-and-conquer technique. For example, it gives the individual the "freedom" to stand alone as a powerless and inevitably naïve consumer against massive organizations like ITT, the federal government, General Motors, and so on. Cooperation, organization, and coordination are necessary for human survival, and individualism is a romantic denial of this—a denial that leads to still larger and more impersonal organizations. In the end it's the huge bureaucracies that derive the "benefits" of individualism. Their life is made easier when they can say, "Go do your own thing; we'll mind the store."

The Automobile

But institutions don't just *express* motivations, they *create* them as well. The automobile is a good example of this—one that only became clear to me when I began to ride a bike. Riding around the city I had the overwhelming feeling that automobile drivers were deranged—the victims of some kind of mass hysteria. I was frightened by their competitiveness, irritability, and screeching impatience—hitting highway speeds to cross a parking lot. It was always particularly bad around five o'clock. But then I discovered that when I got behind the wheel of a car again myself, I was behaving exactly the same way. Perhaps this is why I had never noticed the madness before, in thirty years of driving.

The reasons are simple. Automobiles confer great power and high expectations, and then frustrate them. Their numbers create scarcity, and hence to drive in traffic is to enter a competitive system. Everyone knows this, and many people (myself included) claim to enjoy the struggle. Yet the overwhelming im-

pression I receive when I'm "outside the system" is not enjoyment but rage—a universal, bitter rage. I think it became
noticeable to me by contrast with my own quite different mood
riding a bike, which was generally happy and serene, despite the
considerable dangers of being around these madmen in anything
smaller than a car. My bike trips were quick, pleasurable, and
always took exactly as long as I expected.

I'm now more aware of my own tension when I'm behind the
wheel. A car is built to move very fast. It's expensive and uses a
lot of energy. It should do better than a bicycle, but in the city it
doesn't. In a car one feels that normality is to be moving at least
40 miles per hour—anything less is experienced as abnormal, as
a flaw in the system. Tension accumulates imperceptibly in response to each obstruction. Given the slightest delay or urgency,
drivers tend to become intolerant of all obstacles—they struggle
with destructive impulses. The attitudes of American soldiers
and fliers toward the Vietnamese owed much to the teachings of
the automobile: there was the same feeling that human obstacles
to the smooth movement of the American war machine had no
right to exist. The impulse to run over them, however, was in
that case unchecked.

The automobile, in other words, tends by its very nature to
make people competitive, arrogant, ruthless, and irascible. *It
cannot fulfill the dreams of power and speed that it arouses because it contains within itself no system for coordinating its
movements with those of others.* The automobile is thus the perfect symbol of American life, the ultimate expression of doing
your own thing.

CHANGE AND STABILITY

Another paradox of social change is that those who most actively
seek it are often looking for some kind of stability—a stability
they think the change will bring about. But change is as hard to
stop as it is to start.

All this is particularly true today, since we live under conditions of chronic change brought on by technology. The most real

change we could achieve would be one that would give us a holiday from our frantic and desperate pursuit of the future. We're in the bizarre situation where our radicals want stability and simplicity, while our conservatives dote on chronic change; the old culture worships novelty, while the new culture wants to revive the present and the past. The old culture systematically invalidates experience—turns its back on age and on familial and community ties; while the new culture is preoccupied with tradition, with community, with relationships—things that would reinstate the validity of accumulated wisdom.

Thus we're confronted with the paradox of trying to build a future that doesn't always look to the future. We need desperately to find a social change mechanism that's self-extinguishing. Revolutionaries always assume that change and fascination with the future will cease once the golden day arrives, but they seldom manage even to slow it down.

Changing our culture will require participation by everyone. Children's crusades are fetching and romantic, but they always end with either the death of the children or the absent-minded evaporation of the army. Older people are needed to understand where the present connects with the past (for only the combinations change, the elements are deathless). Otherwise the society we build will have the same defects as the old one.

It might be objected at this point that I'm taking a lot for granted when I speak of dismantling the old culture and nurturing a new one of some kind. How do we know it wouldn't be worse? Isn't that the same faith in the future that caused all our problems in the first place?

My main argument for rejecting the old culture is that as it struggles more and more violently to maintain itself, it's less and less able to hide its fundamental antipathy to human life and human satisfaction. It spends hundreds of billions of dollars to find ways of killing more efficiently, but almost nothing to enhance the joys of living. However familiar and comfortable it may seem, the old culture threatens to destroy us, like a trusted relative gone berserk so gradually that we're able to pretend he hasn't changed.

But what can we cling to—what stability is there in our chaotic environment if we abandon the premises on which the old culture is based? To this I would answer that it's just those premises that have created our chaotic environment. I recognize the desperate longing in America for stability, for some fixed reference point when all else is swirling about in endless flux. But to cling to old-culture premises is the act of a hopeless addict who, when his increasingly expensive habit has destroyed everything else in his life, embraces his destroyer more fervently than ever. The radical change I'm suggesting here is only the revival of stability itself.

Specifically, I'm suggesting that we reverse our old pattern of technological radicalism and social conservatism. Like most old-culture premises, this is built on a self-deception: we pretend that we can actually achieve social stability this way—that technological change can be confined within its own sphere. Yet obviously this is not so. Technological instability creates social instability as well, and we lose both ways. Radical social change *has* occurred within the old culture, but unplanned and unheralded. The changes embodied in the new culture are changes that at least some people want, while the changes that have occurred under the old culture were desired by no one. They weren't even foreseen. They just happened, and people tried to build a social structure around them; but it's always been a little like building sand castles in heavy surf, and we've become a dangerously irritable people in the attempt. We have given technology *carte blanche*, much in the way Congress has always, in the past, given automatic approval to defense budgets, resulting in the most gigantic graft in history.

How long is it since anyone has said: "This is a pernicious invention which will bring more misery than happiness to humanity"? One reason no one does is the sheer difficulty of figuring out equivalents in a complex society. If we could keep our heads aware of the fact that all energy is part of a common pool, we might decide, for example, that getting an instant picture on our TV set instead of waiting a few seconds wasn't worth destroying a brook; or that being able to sit on a big machine while it raked a few leaves, instead of doing it by hand, wasn't worth polluting a beach.

Who would dare defend even a small fraction of the technological innovations of the past century *in terms of human satisfaction?* The problem is that technology, industrialism, and capitalism have always been evaluated in their own terms. But it's as absurd to judge capitalism by the wealth it produces, or technology by the number of inventions it generates, as it would be to judge a bureaucracy by the amount of paper it uses. We need to find ways of appraising these systems with criteria that are truly independent of the systems themselves. We need to develop a human-value index—one that assesses the worth of an invention or a system or a product in terms of its total impact on human life; in terms of ends rather than means. We would then judge the achievements of medicine not by the hours of prolonged (and often comatose) life, or the volume of drugs sold, but by the overall increase (or decrease) in human beings feeling healthy. We would judge city planning and housing programs not by the number of bodies incarcerated in a given location, or the number of millions given to contractors, but by the extent to which people take joy in their surroundings. We would judge the worth of an industrial firm not by its profits, or the number of widgets manufactured, or how distended the organization has become, but by how much pleasure or satisfaction has been given to people. It's significant that we appraise a nation today in terms of its gross national product a phrase whose connotations speak for themselves.

Freud suggested back in 1930 that the much-touted benefits of technology were "cheap pleasures," equal to the enjoyment obtained by "sticking one's bare leg outside the bedclothes on a cold winter's night and then drawing it in again." "If there were no railway to make light of distances," he pointed out, "my child would never have left home and I should not need the telephone to hear his voice."[2] Each technological "advance" is heralded as one that will solve the problems created by its predecessors. None have done so, however—they have only created new ones.

Heroin was first introduced into this country as a heaven-sent cure for morphine addicts, a routine now being replayed with methadone. Technological progress follows the same model: we have been continually misled into supporting a larger and larger technological habit.

A *Cheap Pleasure*

Anyone can think of examples—the relationship between cars, roads, and cities provides a great many—but my favorite illustration concerns an invention which happily seems never to have gotten off the drawing board. I quote from a newspaper article: "How would you like to have your very own flying saucer? One that you could park in the garage, take off and land in your own driveway or office parking lot. . . . Within the next few years you may own and fly just such an unusual aircraft and consider it as common as driving the family automobile. . . ." The writer goes on to describe a newly invented vertical-take-off aircraft which will cost no more to own and operate than a sports car and is just as easy to drive. After an enthusiastic description of the design of the craft, he goes on to attribute its development to the inventor's "concern for the fate of the motorist," citing the inability of the highways and city streets to handle the increasing number of automobiles. The inventor claims that his saucer "will help solve some of the big city traffic problems!"[3] Indeed, so confident is he of the public's groveling submission to technological commands that he doesn't even bother to defend this outlandish statement. In fact, he doesn't even believe it himself, since he brazenly predicts that every family in the future will own a car *and* a saucer. He even acknowledges rather flippantly that air traffic might become a difficulty, but suggests that "these are not his problems," since he is "only the inventor." One is reminded of Tom Lehrer's song about the rocket scientist:

> *"Once they are up who cares where they come down: That's not my department," says Werner Von Braun.*

The inventor goes on to say that his invention would be useful in military operations (such as machine-gunning oriental farmers and gassing dissenters, functions now performed by the helicopter), and in spraying poisons on our crops. His remarks betray an awareness of the rule in our society that while those who kill or make wretched a single person are severely punished, those (heads of state, inventors, manufacturers) who are responsible for the death, mutilation, or general wretchedness of thousands or millions are rewarded with fame, riches, and prizes. The old culture's rules speak very clearly on this: if you're going to rob, rob big; if you're going to kill, kill big.

How can we account for the lack of public resistance to this kind of arrogance? Why does the consumer comply so abjectly with each technological whim? Is the man in the street so punch-drunk with technological propaganda that he could conceive of such a saucer as a solution to *any* problem? Could he ignore the horror of an invention that would blot out the sky, increase our already intense noise level, pollute the air further, facilitate crime immeasurably, and cause hundreds of thousands of horrible accidents (since translating our highway death toll to the air requires the addition of bystanders, walking about the city, sitting in their yards, or sleeping in their beds)? Is the American public so insane or obtuse as to relish the prospect of the sky being as filled with motorized vehicles as the ground is now?

It's conceivable, since Americans are trained by the advertising media to identify immediately with the person who actually uses a new product. So if a man thinks of a saucer, he'll imagine himself inside it, flying about and having fun. He won't think of himself trying to sleep and having other people roar by his window. Nor will he think of himself trying to enjoy peace and quiet in the country with others flying above. Nor will he even think of the other fliers colliding with him as they all crowd into the city. The American, in fact, never thinks of other Americans at all—his most characteristic trait is that he imagines himself to be alone on the continent.

Furthermore, Americans are always hung over from some blow dealt them by their technological environment and are always looking for a fix—for some pleasurable escape from what technology has created. The automobile, for example, did more than anything else to destroy community life in America. It segmented the various parts of the community and scattered them about so that they became unfamiliar with each other. It isolated travelers and decoordinated the movement of people from one place to another. It isolated and shrank living units to the point where the skills required for informal cooperation among large groups of people atrophied and were lost. As the community became a less and less satisfying and pleasurable place to be, people more and more took to their automobiles as an escape from it. This in turn crowded the roads more, which generated more road-building, which destroyed more communities, and so on.

The "saucers" would carry this process even further. People would take to their saucers to escape the hell of a saucer-filled environment, and the more they did, the more unbearable that hell would become. Each new invention is itself a refuge from the misery it creates—a new hero, a new heroin.

An Expensive Habit

How far can it go? What new inventions will be offered the staggering American to help him blow up his life? Will he finally flee to outer space, leaving the nest he has so industriously fouled behind him forever? Can he really find some means to propel himself so fast that he'll escape his own inventive destructiveness? Is the man in orbit—the true Nowhere Man, whirling about in his metal womb unable to encounter anyone or anything—the destiny of all Americans?

If we fail to kick this habit, we may retain our culture and lose our lives. One often hears old-culture people say, "What will you put in its place?" But what does a surgeon put in the place of a malignant tumor? What does a policeman put in the place of a traffic jam? What do we put in the place of war when peace is declared? The question assumes that what exists is safe and tolerable.

Yet it's true that the old culture is highly dependent on the palliatives that its own pathology necessitates. "Without all these props, wires, crutches, and pills," people ask, "how can I function? If you take away my gas mask, how can I breathe this polluted air? How will I get to the hospital without the automobile that has made me unfit to walk?" These questions are serious, since one cannot in fact live comfortably in our society without these props until radical changes have been made. Transitions are always fraught with risk and discomfort and insecurity, but we don't enjoy the luxury of postponement.

Our servility toward technology, however, is no more dangerous than our exaggerated moral commitment to striving and personal achievement. The mechanized disaster that surrounds us is in no small part a result of our having deluded ourselves that a motley scramble of people trying to get the better of one another is socially useful, instead of something to be avoided at all costs. It has taken us a long time to realize that always trying to surpass people might be sick, and trying to enjoy and cooperate with people healthy, rather than the other way around.

The need to triumph over others and the tendency to prostrate ourselves before technology are in fact closely related. We turn continually to technology to save us from having to cooperate with one another. Technology, meanwhile, helps preserve the competitiveness and render it ever more frantic, thus making cooperation at once more urgent and more difficult.

The essentially ridiculous premises of a competitive society are masked not only by technology, but also by the ability men have to compartmentalize their thinking about economics. Since they're more interested in achievement than in satisfaction, they always think of themselves first as producers and only second as consumers. They talk of the "beleaguered consumer" as if this referred to some befuddled group of little old ladies. Since men have traditionally dominated production, and women, consumption, the man who produces shoddy merchandise could blame his wife for being incompetent enough to purchase it for him. Men have insulated themselves from having to deal with the consequences of their behavior.

What all of our complex language about money, markets, and profits tends to hide is the fact that when the whole circuitous process has run its course, we're producing for our own consumption. When I exploit and manipulate others, I'm also exploiting and manipulating myself. The needs I generate create a treadmill that I myself will walk on. It's true that if I manufacture shoddy goods, create artificial needs, and sell food that looks good but is tasteless or contaminated, I'll make money. But what can I do with this money? I can buy shoddy goods and poisoned food, and satisfy ersatz needs. Our refusal to recognize our common economic destiny leads to the myth that if we all overcharge each other, we'll be better off.

This self-delusion is particularly impressive when it comes to issues of health and safety. Are executives who live in cities immune to the air pollution caused by their own factories? Or do they all live in exurbia? And have oil company executives given up ocean beaches as places of recreation? Do they all vacation at mountain lakes? Do automobile manufacturers share a secret gas mask for filtering carbon monoxide out of the air? Are the families of chemical manufacturers and farming tycoons immune to insecticides?

To some extent wealth does purchase immunity from the effects of the crimes perpetrated to obtain it. But in many cases the effects cannot be escaped even by those who caused them. When a tanker flushes its tanks at sea, or an offshore well springs a leak, the oil and tar will wash up on the most exclusive beach as well as the public one. The food or drug executive can't tell his wife not to purchase his own product, since he knows his competitors share the same inadequate controls. We cannot begin to understand the irresponsibility of corporations until we recognize that it includes and *assumes* a willingness on the part of corporate leaders to endanger themselves and their families for the shortrun profit of the corporation. Men have always subordinated human values to the mechanisms they create. They have always been able to invest their egos in organizations that they then view as having independent life and superior worth. Man-as-thing (producer) can then enslave man-as-person (consumer), since his narcissism is most fully bound up in his "success" as a producer. What he overlooks, of course, is that his success as a producer may bring about his death as a consumer.

CHANGE AND INCENTIVES

One might object that this discussion puts too much emphasis on individual motivation. We can't expect, after all, that everyone will get up one morning suddenly resolved to act differently, and thus miraculously change the society. Competitive environments are hard to modify, since whoever takes the first step is very likely to go under. "The system" is certainly a reality, even if it's made up of fictions.

Any action program must consist of two parts: (1) a long-term thrust at changing motivation and (2) an immediate attempt to transform institutions. As the motivational underpinnings of the society change (and they're already changing) new institutions will emerge. But so long as the old institutions maintain their present form, they will tend to overpower and corrupt the new ones. They must be changed so that they no longer *reward* greed and arrogance. In other words, the incentive structure must be changed as the motivations shift.

Imagine that we're all marooned on a large and inescapable boat, sailing in a once ample but now rapidly shrinking lake. For generations we've been preoccupied with finding ways to make the boat sail faster around the lake. But now we find we've been all too successful, for the lake gets smaller and smaller and the boat goes faster and faster. Some people are saying that since the lake is about to disappear we must develop a new way of life— to learn to live on land. They say that going in circles on a little lake is absurd anyway. Others cling to the old ways and say that living on land is immoral. Then there's a middle-of-the-road group that says living on the lake is best, but that perhaps we'd better slow down before we smash to pieces on the ever-nearer rocks around and below us.

Now if it's true that the lake is disappearing, the radicals must not only prepare themselves and convert others, but must also train the captain and crew to navigate on land. And the middle-of-the-roaders must not only find ways to slow the boat down, but need also to find some way to attach wheels to its bottom. Putting wheels on the boat is what I mean by reforming the incentive structure. It's a way of softening the collision between old and new.

Let me give a concrete example of adjusting institutions to match changes in motivation. It seems clear that fewer people today are interested in careers of personal aggrandizement compared with twenty years ago. Far more want to devote themselves to something socially useful. This is surely a beneficial shift in emphasis—we perhaps don't need as many people as we once did to enrich themselves at our expense, and we have no place to put the overpriced junk we already have. But our old-culture institutions continually put obstacles in the path of this shift. There are few jobs available for people who want to be socially useful, since as a nation we don't like to spend money on things like that. Such people have swelled the ranks of the un-employed. Some are living marginal existences in rural areas, others are "getting by" in the cities, living on welfare or selling drugs, often doing volunteer work on the side. At a conservative estimate there are probably a million men and women in their twenties and thirties who would happily work long hours doing what most needs to be done if they were paid something for it. This is a criminal waste of a precious national resource.

Furthermore, those who seek to provide services are often prevented by established members of the professions—such as doctors, teachers, and social workers—since the principle be-hind any professional organization is (a) to restrict membership and (b) to provide minimum service at maximum cost. A profes-sion, after all, is simply a closed-shop union, set up in opposition to the consumer instead of to management. Requirements for credentials are designed not to create excellence but merely to limit membership and keep prices up. In the fields I know best—university teaching and psychotherapy—the kind of train-ing that leads to proper credentials is often actually detrimen-tal—at best it's irrelevant. There are equal numbers of excellent therapists, for example, among the credentialed and the uncre-dentialed, and equal numbers of dangerous incompetents. The same may be said of teachers. The only "degree" worth anything is the recommendation of a satisfied customer.

The Greed Prize

The Internal Revenue Service is another obstructive institution. The whole fabric of income tax regulations is woven around the principle of rewarding single-minded devotion to self-aggrandizement. If you spent all your money protecting, main-

taining, or trying to increase your income, you would theoreti-
cally pay no tax whatever. The tax structure rewards the money-
grubber, the wheeler-dealer, and punishes the person who
simply provides a service and is paid something for it. The person who devotes his life to making money is rewarded by the
United States Government with tax loopholes, while the person
who devotes her life to service picks up the check.

We need to reverse these incentives. We need to reward
everyone *except* the money-hungry—to reward those who are
helping others rather than themselves. Actually, this could be
done very easily just by eliminating the whole absurd structure
of deductions, exemptions, and allowances, and thus taxing the
rich and avaricious instead of the poor and altruistic. This would
have other advantages as well: discouraging overpopulation and
home ownership, and saving billions of hours of senseless and
unrewarding clerical labor. Tax loopholes cost the federal government almost 60 *billion* dollars a year, and more than half of this
goes into the pockets of the rich—the one-seventh of the population with the highest incomes.

Reforming the priorities involved in the disbursement of
federal funds would also help. At present, most of the federal
budget is devoted to life-destroying activities, less than half to
life-enhancing ones. Most government spending, furthermore,
subsidizes the rich: defense spending subsidizes war contractors,
foreign aid subsidizes exporters, the farm program subsidizes
rich farmers, highway and urban redevelopment programs subsidize building contractors, medical programs subsidize doctors
and drug companies, and so on. In order to subsidize the poor as
effectively as we subsidize the rich we would have to guarantee
employment for all.

Our goal is not to make money-grubbers out of those who
aren't, but rather to restore money to its rightful place as a medium of exchange—to reduce the role of money as an instrument
of vanity. As it is now, those with the greatest need for self-
aggrandizement can amass large surpluses, and the government
tends to reward and encourage this. The shortages thereby created tend to make it difficult for people who are less self-seeking
to maintain a casual attitude toward money. The poor, mean-

while, are thrown into such an acute state of deprivation that money comes to overshadow all other goals. Since familiarity with money often breeds contempt, whatever we can do to equalize the distribution of wealth will help create disinterest. This will leave only the most egocentric still mad for money—indeed, they'll be worse than ever, since they will have been deprived of their surplus millions or of the opportunity of amassing them, and will have to look elsewhere for the means of satisfying their vanity. Perhaps they'll seek it through the exercise of power—becoming generals or teachers or doctors; perhaps through fame, becoming writers or artists or scholars. Like the poor, the vain are always with us, but for the ordinary person money would then be used merely to obtain goods or services.

Such a profound transformation is not likely to occur soon. But it's interesting that a reversal of the incentive structure is what is most feared by critics of such plans as the negative income tax. Why would people want to work and strive, they ask, if they could get all they wanted to eat without it? Why would they be willing to sell out their friends, sacrifice family ties, cheat and swindle themselves and everyone else, and disregard community needs, if they could obtain goods and services without doing these things? "They would have to be sick," we hear someone say, and this is the correct answer. Only the sick would do it—those who today when they have a hundred thousand dollars still try to make more. *But the non-sick would be free from the obligation to behave as if they were sick—an obligation our society now imposes.*

A Ransom for America

It would not be so difficult, if these proposals were carried out, for Americans to be motivated by something other than greed. People engaged in helping others, in making communities viable, in making the environment more attractive, would be able to live more comfortably if they wanted. Some people would of course do nothing at all but amuse themselves, like the idle rich, and this seems to disturb people: "subsidized idleness," they call it, as if to discredit it. Yet I personally would far rather pay people *not* to make nerve gas than pay them to make it; pay them *not* to pollute the environment than pay them to do it; pay them *not* to inundate us with instant junk than pay them to do it; pay them *not* to swindle us than pay them to do it; pay them *not* to kill peasants than pay them to do it; pay them *not* to be dictators than pay them to do it; pay them *not* to replace communities with highways than pay them to do it, and so on. One thing must be said for idleness: it keeps people from doing the Devil's work. The great villains of history were busy men, since great crimes and slaughters require industry and dedication.

What I'm suggesting, then, is that every institution, every program in our society should be examined to see whether it encourages people to serve the community or their own egos. This is especially true of the family. For no matter how much we try to eliminate egocentric incentives from other institutions, they will re-emerge if we don't devote some attention to reforming the family patterns that produce egoism in children.

Some people may feel that this is already happening. The "sexual revolution" promises to eliminate one source of scarcity psychology, and this liberalization of sexual norms has led to a more generalized movement toward the liberation of women (perhaps because historically, sexual restrictions have always been imposed primarily on women). Mothers of the future should therefore be far less driven to flood their male children with frustrated longings and resentments. Living fuller, less constricted lives themselves, they will have less need to invest their sons with Oedipally-tinged ambition.

The Well-Worn Path

There is certainly no cause to lie back and wait for the better day, however, for these changes are far from automatic, and youthful enthusiasm is no match for old-culture institutions. There are three gates into the mainstream of our society: employment, marriage, and parenthood, and each is a more powerful instrument of old-culture seduction than the last. Indeed, old-culture adults count heavily on this triple threat to force young people to abandon their new-culture lifestyles ("wait until you have to raise a family"), and are very disappointed when the threat fails ("you have to grow up *sometime*").

But usually their expectations are confirmed—not because there's anything inherently mature about competing with your neighbor, but because the new culture has made few inroads into the structure of adult life. Young people with new-culture ideas find themselves leaving environments in which their attitudes were widely shared, and moving into one in which they are isolated, surrounded, and shunted onto a series of conveyor belts that carry them into the old culture with an inevitable logic that can be resisted only with deliberate and perpetual effort. Often they know this and fear it. They dread becoming like their parents but can't see how to avoid it. It's as if they had come to the edge of a dense, overgrown forest, penetrable only by a series of smooth, easily traversed paths, all of which have signs saying TO THE QUICKSAND.

Work is the first seduction, although with great struggles, floundering, and anxiety, many people are managing to carve out lives for themselves that don't commit them fully to the old culture. Graduate training of any kind is usually the most fatal of all pathways, but even in the professions some people have been able to retain new-culture values, and every field has its small but indestructible radical wing—largely unemployed.

Fear of marriage and bad marital relationships is strong, but young people still marry in droves; while parenthood is often not feared at all, although it is clearly the most dangerous. For it was parenthood that played the largest part in the corruption of their own parents: "For the children" is second only to "For God and country" as a rallying cry for public atrocities. The new parents may interpret the slogan in less materialistic terms, but the old culture and the new share the same child-oriented attitude. This creates many pitfalls for unwary young parents, since the old culture has a built-in system of escalating choice-points to translate love for children into old-culture practices. The minute the parents decide they want their child to have some green grass to run about in, or a school that isn't taught by rigid, authoritarian teachers, they're apt to discover that they have eaten a piece of the gingerbread house and are no longer free.

Even here, of course, there are solutions, but less thought and attention have been given to them by the young. They imagine, like every fool who ever had children, that their own experiences as children will protect them against their own parents' errors. People in our society are particularly blind to the overwhelming force of identification, and they're also peculiarly unprepared, by the insulation of their youth culture, for its sudden onset. In more traditional societies everyone realizes that upon becoming parents they will tend automatically to mimic their own parents' behavior, but in our society this comes as a shock, and is often not even realized when it happens. If the young do reproduce the behavior of their child-oriented parents, there is danger that the timely provision of a target for their feelings of responsibility will drain off their concern for the community.

It's difficult, in other words, not to repeat patterns that are as deeply rooted in primary emotional experiences as these are, particularly when one is unprepared. The new parents may not be as absorbed in material possessions as their own parents were—they may channel their parental vanity into different spheres, pushing their children to be brilliant artists, thinkers, and performers—but the hard narcissistic core on which the old culture was based will not be dissolved until the parent-child relationship itself is de-intensified, and this is precisely where the younger generation is likely to be most inadequate. While the exterior of the old culture is being weakened, its nucleus is in danger of being transferred, not only intact, but strengthened—like a bacterial strain resistant to drugs—to the new.

Being child-oriented doesn't in itself produce a narcissistic personality. It's when the parent turns to the child as a substitute for satisfactions missing in his or her own life that the child becomes vain, ambitious, hungry for glory. Both the likelihood and the intensity of this pattern are increased when the family is a small, isolated unit and the child has few important ties with other adults. Our society has from the beginning, and increasingly with each generation, tended to produce "Oedipal" children. New-culture youth often want desperately to build a co-operative, communal world, but they are in some ways the least likely people in the world to be able to do it, or to produce children that could do it. They can't break the Oedipal pattern alone because they're even more enmeshed in it than were their parents.

Breaking the pattern means establishing communities in which (a) children are not brought up exclusively by their parents, (b) parents have lives of their own and do not live vicariously through their children; hence (c) life is lived for the present, not the future, and hence (d) people of all ages and both sexes participate in the total life of the community. This is one reason why I'm not particularly interested in any supposedly "new" development that doesn't include changes in the relations between the sexes. No organization, collective, or political movement dominated in the usual way by men will ever lead to any significant social change. No matter how radical they may appear on the surface, they mean little more than going back and rerunning our same old tape from some earlier starting point. The significant innovations today are being made by women— most men are merely running minor variations on familiar ego-themes. Since men tend to do what they do for notoriety, however, this may not be widely apparent for a long time to come.

CHANGE AND POWER

Americans have always thought that change could occur without pain. This pleasant idea springs from a confusion between change (the alteration of a pattern) and novelty (the rotation of stimuli within a pattern). Americans talk about social change as if it involved nothing more than rearranging the contents of a display window. But real change is difficult and painful, which perhaps explains why Americans have abandoned all responsibility for initiating it to technology and the rotation of generations.

Severe illness sometimes leads to great strengths, madness to inspiration, and decay to new growth. Building on its tortured history, the United States could become the center of the most beautiful and exciting culture the world has ever known. We have always been big, and have done things in big ways. Having lately become in many ways the worst of societies we could just as easily become the best. No society, after all, has ever solved the problems that now confront us. Potentiality has always been our most attractive characteristic, which is one reason we've always been so reluctant to commit ourselves to anything. But perhaps the time has come to make that commitment to abandon our adolescent dreams of limitlessness and demonstrate that we can create some kind of palatable society. America is like a student who is proud of having survived school without serious work, and likes to imagine that if he really put any effort into it, he could achieve everything, but is unwilling to endanger so lovely a dream by making an actual commitment to a single project. Yet we have a great deal to work with. Our democratic institutions—tattered as they are—are a great resource, although we've been too caught up in individualistic dreams of glory to make much use of it. Despite our lack of community and our chronic flirtations with authoritarianism, no society of comparable size is as responsive to popular sentiment.

The fundamental political goal of the new culture is the diffusion of power, just as its basic economic goal is the diffusion of wealth. Marxists want to transfer concentrated power into the hands of revolutionaries first, in order to secure the diffusion of wealth. In the United States, however, the diffusion of wealth is a more easily attained goal than the diffusion of power, so that it becomes more important to ensure the latter, and to be skeptical of its postponement.

Media

Activists have often achieved some dispersion of power on a local scale merely by unmasking, exposing, or threatening to expose those at the centers of power. The ability to maintain a concentration of power depends upon the ability to maintain and enforce secrecy, and dispersal of power tends to follow automatically upon the breakdown of such secrecy. Old-culture leaders are peculiarly vulnerable on this point because they're not sensitive to the nature of mass media. They think in terms of news management and press releases and public statements—of *controlling* the media in the old-fashioned propagandistic sense. Traditional Marxists share their views, and devote their energies to worrying about the fact that all news media are controlled by a relatively small number of wealthy and conservative men. New-culture activists, on the other hand, are attuned to the media.

They know that *the media are inherently stimulus-hungry,* and must by their very nature seek exposure and drama. They know that a crowd is more interesting than a press conference, a march more interesting than a speech. Successful use of television to-day requires an improvisational looseness and informality that old-culture leaders lack. Their carefully managed statements be-come too obviously hollow with repetition, their pomposity too easily punctured by an awkward incident, their lies too recently stated and well-remembered to be ignored. It seems astonishing to us when statesmen and generals constantly put themselves in the position of saying, in effect, "Well, I lied to you before, but this time I'm telling the truth." But prior to television it was quite possible to assume that the mass of the population was sub-stantially without memory.

I'm suggesting that the diffusion of power could occur with little change in the *formal* machinery of government, which, after all, lends itself to a wide range of political types. Yet it can't be denied that at present the power of the executive is over-whelming and could in theory easily crush any dissident group. I say "in theory," because if all the other parts of the system are working properly, any such repression will automatically invali-date the presidency and bring it down.

Ever-Present Dangers

This fact creates an odd dilemma: awareness of the power of the democratic process can lead to the apathy that will destroy it, while belief in the theoretical power of the executive, though simple-minded, creates the vigilance necessary to curb it. In the first edition of this book I made dire comments about how easy a right-wing takeover would be—once again, in theory. I pointed out that old-culture conservatives had access to our gigantic ar-senal of weapons and wouldn't hesitate to use them; that the mil-itary, CIA, FBI, and other law enforcement and intelligence groups were veiled in secrecy—much more so than any other segment of the population—and could operate in undercover ways not possible for anyone else; that they were the only people equipped to ferret out and expose such a plot, yet were heavily populated by the right-wingers most likely to participate in it. I also pointed to the fact that the laws and machinery existed to put every leftist or liberal in the country in concentration camps in response to a single executive order. I then suggested that the danger would increase as confrontations between the two cultures increased.

This apparently made a strong impression on many people, for I have frequently been asked about it in the intervening years. My response was always that the danger would lessen every year that it didn't happen. My reasoning was that the existence of constant confrontation and disruption would provoke such a coup, since most Americans just want to go about their business and ignore social problems, and would support anyone who promised, like Hitler, to restore law and order. But the more people got used to the new ideas and embraced the new values, the less such a coup would be possible. I now tend to feel that the right missed its chance. Its most sacred precincts—the CIA, FBI, and Pentagon—have been challenged and in part discredited, while Congress and the media have shown new backbone since Watergate. This was not unexpected, for reasons I'll get to shortly. What I didn't anticipate was that the radical movement would run out of steam and disperse, thus taking the pressure off and allowing people to absorb the ideas at a pace they could handle. I spoke about "easing the transition" from the old culture to the new, but this wasn't exactly what I had in mind.

Yet I want to emphasize that the danger has by no means disappeared. In another period of social turmoil and upheaval, with any lapse in vigilance over those nondemocratic pockets and corners of our government, a few right-wing fanatics could still gain control over enough lethal weaponry to terrorize the country into submission. The more hopeful things I'm about to say should not be taken as cause for relaxation but rather as spurs to even greater activity.

The United States is still in the process of becoming democratized. Slavery, expansion, immigration, industrialism, and war have undercut our efforts to realize our political and social ideals. In some sense we've always been off-balance—always seduced into authoritariansm in the face of some real or imagined social disruption. This isn't because authoritarianism works better or is

good in crises, but because people latch on to it when communication between social groups breaks down.

Why Democracy Survives

Warren Bennis and I pointed out more than a decade ago[4] that
democracy is not a luxury but the most efficient mode of organization under conditions of great complexity and chronic change.
We showed that the "efficiency" Americans usually attribute to
autocratic systems applies only to situations involving simple
routine tasks. Such systems function poorly when the world becomes intricate and shifting. They have an awkward tendency to
run a "tight ship" which nevertheless sinks. Authoritarian leaders
are so bound into their own centrality and power that they always
miss or ignore the important cues that allow democratic systems
to change and adapt. Short-run autocratic "efficiency," in other
words, always leads to long-run collapse because it's too rigid and
its leaders too inaccessible to information from the system's periphery. The corporation manager who through his "efficiency"
saves a hundred thousand dollars in a cost-cutting program that
provokes a million-dollar strike typifies this kind of tunnel vision.

Bennis and I used Hitler as the classic example, but our theo-
ries fit Watergate with some precision. It wasn't the arrogance or
the dirty tricks themselves that did Nixon and his staff in: they
could have owned up to and minimized the burglary immediately and gotten away with it. An admission of "faulty judgment,"
someone resigns, it blows over. It was their out-of-touchness
with the political process that did them all in—their overesti-
mation of their own power and control (exemplified by Nixon's
overweening vanity in not destroying the tapes), their insensitivity to the media, and their blindness to that crucial moment
when closeness to the power center becomes a liability rather
than an asset and it's "every man for himself." Watergate, in other
words, was not an accident but an inevitability—anything could
have triggered it. It follows from a fundamental political law: the
fewer people there are who know what's going on *inside* an office,
the less *they* know about what's going on *outside* of it.

This law is basic to democracy—it's what makes it work. It's a
simple matter of numbers: the fewer there are of you, the fewer

eyes and ears you have, the more narrowly information has to be funneled to you. The despot only gets to hear what he's used to hearing and what he's comfortable hearing, until it's too late.

Unfortunately, for these very reasons, democracy is its own worst press agent. Democracy is a long-range strategy, but because it diffuses decision-making and thereby maximizes small changes—capricious-seeming little adjustments—it feels chaotic, and tends to breed short-run thinking in its constituents. And because its constituents think in short-run terms, they can't appreciate the value of a long-range strategy, and don't see how or why democracy works.

Authoritarianism is the exact opposite—a short-run strategy that fosters long-range thinking. In the false calm of controlled, constricted, and inflexible decision-making, people are encouraged to think in long-range terms. The Third Reich was designed for a "thousand years," but it lasted only twelve.

But although democracy may never be appreciated for what it really is—an intelligent adaptive strategy—Watergate was nonetheless educative. I have had many people congratulate me for predicting it (although I certainly never did) and say they now understood for the first time what Bennis and I were trying to say. It's not just that Watergate alerted people to the enormous temptations to tyranny that abound in the executive branch of our government, nor that it led to other investigations and some minor efforts to curb executive power. What really excited people was some dim realization that the system actually works; that it has an organic life of its own that to some degree is self-corrective; that the people haven't yet managed to give all their power away (not for lack of trying, certainly)—that they still have some and can use it.

Individualism: The Midwife of Fascism

Mostly this has been stimulating to the democratic process, since it creates hope where there was cynicism and despair. Lest it also create complacency, however, I might point out that the temptation to give power away is very strong in Americans because of our individualistic heritage. We're always wanting to give power to leaders so we can do our own thing and not be bothered by the demands of collective commitment. This trend is most exaggerated in universities, where professors are always in a frenzy because some administrator has actually used the power they've abdicated to him. Individualism is so rampant in the university that meetings of any kind are peculiarly unbearable—a group of managers seems in warm and telepathic rapport by comparison. Academics as a group are so alienated from their feelings that they tend to invest them in every trivial procedural question rather than expressing them directly. This makes cooperation virtually impossible, and rather than put up with each other's posing and sermonizing any more than they already have to, they are seldom able to resist putting crucial power into the hands of administrators.

This is a violation of democracy's first rule: never delegate authority upward. It's a rule violated by liberals more than anyone else, since liberals are most uncomfortable with the demands of communal existence. Cooperation is so irksome to individualistic natures that they spend half their political lives giving power to centralized governments and the other half fearing for their personal liberties, without ever considering the contradiction.

I have said that it's important to diffuse power, to detach it from those who hate life and would rather die themselves than see others enjoying it. But this raises an awkward dilemma: in a satisfying society, who else would want power? What but a kind of sickness would drive people to attain such power over others? Wouldn't the sickest people wind up with the most power, even more regularly than they do now? If power is diffused, on the other hand, won't the entire population be corrupted by this sickness?

The answer to this last question is No. Nothing is poisonous if taken in small enough quantities, and the more power is diffused, the more the assumption of power looks like the assumption of responsibility. It's when power is concentrated that the pursuit of it takes on an unhealthy hue. One of the best arguments for participatory democracy is that the alternative to participating in the drudgery of government is being governed by the sick and perverse.

Evil is created by negation—to use the word is to extend its domain. When we reject some human trait, exile it, push it down, wall it up, and scorn it, it becomes malevolent, perverse, vicious. Those who refuse to dirty their hands by exercising power and responsibility for the coordination of community affairs often find themselves murdered in the night by those with dirty hands. Any trait that's harmless when accepted as part of the human condition assumes a poisonous concentration when it isn't. Anger becomes implacable malice, self-respect becomes vanity, the wish for a place becomes individualism, the desire for stability becomes reactionary conservatism, the need for order becomes tyranny, the need for coordination becomes witch-hunting.

One of the oldest and deepest myths in the American psyche is the fantasy of being special—a treat that American society promises, and withholds, more than any society in history. The unstated rider to "do your own thing" is that everybody will watch—that a special superiority will be granted and acknowledged by others. But in a satisfying society this specialness isn't needed, and a satisfying society rests on the recognition that people can and must make demands on one another. Any community worthy of the name—one in which the relationships between people are regulated by the people themselves instead of by machines—would probably seem "totalitarian" to today's youth, not in the sense of having authoritarian leadership, but in the sense of permitting group intrusion into what most Americans consider private. We have long been accustomed to an illusory freedom based on subtle compulsion by technology and bureaucratic institutions. But there is no way for large numbers of people to co-exist without governing and being governed by one another unless they establish machines to do it, at which point they risk losing sight (and understanding) of their interconnectedness—a loss well advanced in our society today. There's something wildly comic about cars stopping and starting in response to a traffic light, for example, but most Americans have lost the capacity to experience it. It seems right and natural for machines to tell us how to relate to each other.

The goal of many early Americans was to create a utopian community, but they became distracted by dreams of personal aggrandizement and found themselves farther and farther from this goal. When we think today of the kind of social compliance that exists in such communities (as well as in the primitive communities we romanticize so much), we shrink in horror. We tell each other chilling stories of people in imagined societies of the future being forced to give up their dreams for the good of the group, or of not being allowed to stand out. But this, in some degree, is just the price one must pay for a tolerable life in a tolerable community. We need to consider this price, to reflect both on its consequences and on the consequences of not paying it. Is an occasional group viciousness really worse than the unfocused universal snarl that has replaced it in our mechanically regulated society? It's the structured narcissism of the old culture that brings down on our heads the evils we detest, and we will only escape those evils when we have abandoned the narcissistic dreams that sustain them.

SEVEN

Your Money or Your Life

You can't always get what you want,
But if you try sometime
You just might find
You get what you need.

JAGGER AND RICHARDS

The strengths of American society lie in its flexibility, its loose-ness, and in the depth with which democratic and equalitarian values are held. Not that these values prevail, not that they ar-en't flouted every day. But occasionally they can be called upon, like sleeping giants, and if awakened long enough, have the power to shift behavior, however gradually. Furthermore, the fact that people feel only loosely tied to one another permits great adaptability to changing conditions.

The weaknesses of our society are tied to its strengths. The flexibility that individualism gives us, for example, is under-mined by the delusions it encourages: that our fates are not in-tertwined, that our neighbor's suffering will not ultimately rub off on us. The traffic jam is a symbol of what this illusory freedom does to us: a whole lot of people pretending they're unrelated to one another and hence frantic with frustration, unable to move.

More and more of our institutions are becoming traffic jams. Our ciries, our economy, our handling of energy, our use of hu-man resources, our ecology, health, education, law and its en-forcement—it's hard to find a sector of society that isn't in this kind of trouble. Does this mean we have to abandon the advan-tages of looseness? In order to adapt, must we give up our ability to adapt?

I find it hard to imagine our individualistic impulses being completely wiped out, but I think we've gone as far as we use-

fully can in the other direction: the only societies I've heard of
that were more individualistic than ours were terminal cases. Yet
there is enough life and creativity in our society to turn this
weakness into a strength. The farther a man wanders, after all,
the more knowledge he brings back when he returns—up to the
point where he wanders too far and disappears forever.

As a society we're at that crucial point now—we can either
use what we've learned from our wanderings and ordeals to enrich our community, or disintegrate altogether. We can recognize
our interdependence and build a society on it or wander off into
oblivion. But precisely because we've been blind to our common
fate for so long, any society we create together will have an organic fluidity that exists nowhere else in the world.

Far from trying to destroy our individualistic heritage I'm
only trying to collect on it. We're not going to stop being who we
are—impatient of constraint, leery of group pressure, prone to
doing our own thing—but we need to put aside our delusions.
To believe that our fortunes aren't tied to everything that lives is
a stupid and costly error.

The problems we struggle with in our society can be traced in
part to the idea that encouraging greed is good for people—that
my trying to take your money away from you will not only make
me rich but you as well. This is a surprising idea coming from a
culture based on scarcity, and it suggests to me that my remarks
in chapter 5 about the two cultures need a little elaboration.

While the new culture believes that the wherewithal to gratify
human needs is infinite, there's a hidden clause: "provided one
is enlightened enough to realize it." We would have to detach
ourselves from material goods, neurotic dreams, and all kinds of
learned addictions for this statement to be literally true. Otherwise, it gets degraded into trying to stay high all the time and
ripping off your friends. The old culture, on the other hand, believes that although emotional gratification is inherently scarce,
an infinite amount of material wealth can be created if everyone
starts grabbing for their piece of the pie.

The grain of truth in this pleasant fiction is that money does
motivate people to expend energy, and when people release energy, something is created. The new culture argues that the best
way to spend energy is on relationships—that you get what you

give, love breeding love, energy breeding more energy. The old
culture opts for matter rather than energy: buildings, machines,
material possessions of all kinds.

Money Talks, But What Does It Say?

There are two basic flaws in basing a society on greed. The first
is that while money motivates people to expend energy, it isn't
the only thing that does so: people will work for friendship, for
love, for the privilege of being part of a working group, for the
enjoyment of service, for beliefs and convictions, or just to make
their environment more attractive. The country is full of people
working "for nothing" (an American term meaning "for no
money"). Nor is this "intangible." When an artist, for example,
fixes up a loft for her own enjoyment, property values and rents
go up. Scott Burns points out that if the unpaid household labor
performed by both women and men in the United States were
given *minimum* remuneration, it would equal the entire amount
paid out in wages and salaries by all the corporations in the
United States.[1] So much for money as the great motivator.

The second problem with money is that it always creates
added motives you don't want. Money (we have to keep remind-
ing ourselves) is a medium of exchange—a *means*. When we use
it as an inducement, people often forget what they wanted it for
and it becomes an instrument of personal or collective narcissism. Instead of purchasing gratifications with it, they accumulate it to demonstrate their worth. It isn't money that runs our
economy but vanity. A corporation may justify its existence in
terms of producing needed goods, but in fact only a small portion
of its resources are devoted to this end. Because it wants to be
bigger than any other corporation, it must expand, go national,
and hence spend enormous sums on advertising, public relations, a national distribution system, and a vast administrative
structure to hold the whole thing together.

Money does motivate people but mainly for the wrong things.
That is to say, *money motivates people to do what they wouldn't
otherwise want to do.* But who is this "people"? To paraphrase
Pogo, we have met the hired help and they is us. *Why do we
want to bribe ourselves to do what we don't want to do?*

Taking Off the Blinders

Perhaps we need to look at our whole economic edifice with naíve eyes to recapture our perspective and see if the emperor really has any clothes on or not. I realize that what I've said so far may seem absurd to economists and sociologists, who take for granted that all of us are alienated from ourselves and must look at the world from a schizoid perspective. Instead of starting with people and working from what people want, economists like to start with tasks: improving the condition of the corporations, or the market, or the interest rate, or the GNP. They talk of creating jobs, markets, demand. Economists assume that jobs must be created even for things that don't need to be done so people can have money to spend on things they don't need. And to get people to buy things they don't need, we create a huge industry to get them to want them. Meanwhile, the things people really need—food, shelter, safety, health, a pleasant environment— they can't afford.

Money is supposed to be a tool—a means to some other end. But economists say things like, "What will be the effect of such-and-such a change on the Gross National Product, on employment, on the interest rate, on the stock market, on inventories, on new plant investment," and so on. These are only measurements of means—what is the goal? *What do we want to do with our work and our resources?* Just make jobs? Just make money? A job *does* something. Money *buys* something. We keep forgetting what it is we *want*—what kind of environment, what kind of life. Do we work only to make more work, and get money only to accumulate more money?

There's something demented about the way we discuss our economy. We put energy into make-work projects and don't get done what we need most. We spend half our days saving steps, the other half jogging, half our energies inventing labor-saving devices, the other half inventing needs that will create jobs. Scott Burns points out that if manufacturers improved the quality of their products so that they lasted twice as long, corporate profits and employment rates—our two main economic indicators— would fall to levels we consider disastrous.[2] But if a calamity can be brought about simply by people doing their jobs well, then our economic thinking is obviously addle-brained.

The last thing we ever think about economically is what we need or want. People often oppose efforts to cut down air or water pollution, or reduce solid waste, on the grounds that polluting the environment keeps people employed. But cleaning up the environment would also "keep people employed," not in make-work, but in something that would give pleasure to people. In the long run no sound economy can be based on useless or destructive labor. The troubles we're experiencing didn't arise because someone made the wrong economic prediction, or used the wrong economic indicator or the wrong theory of corporate investment. They arose because we've been using our energies mindlessly for decades; we've put our labor and resources into activities that have brought us nothing back.

Much of our energy, for example, has gone into war since 1940. War is supposed to stimulate capitalist economies because it "motivates people" to work hard, expand plants, and produce goods that will be instantly destroyed—a bottomless market, the economists' ideal. What's ignored in this convoluted reasoning is that While pouring energy down a rathole is stimulating for a while, it tends ultimately to impoverish. While we've been beefing up our military position, all our basic services have suffered. We may not know that we're exhausted and impoverished but we are. No nation that can't afford to feed, clothe, or shelter its poor, heat its homes, heal the sick, educate the young, or maintain a pleasing environment can call itself wealthy. All we've managed to do is give money to a few hundred thousand *individuals* who are wealthy enough to buy out of all these difficulties. *They* can feel that we're a wealthy nation, and since they command disproportionate power in the land, their somewhat distorted viewpoint often prevails. This is why government economists can speak of economic recovery while the poor are progressively losing jobs and income during severe inflation. The poor scarcely enter into their calculations. But at present, power is the only department in which we hold clear superiority in the world. In most others—health, literacy, nutrition, and so on—we fall below other highly civilized nations. But we can kill and destroy better than any other country, with the possible exception of the Soviet Union.

The issue of taxing corporations provides another example of classic economic reasoning. Between 1955 and 1970 corporate income taxes dropped from 20 percent to 12 percent of federal collections (while payroll taxes almost doubled).[3] This essentially takes money away from services that would benefit everyone and gives it to the wealthy. The argument usually offered to support this theft is that accumulated wealth spurs development, while taxing corporations discourages expansion. In the last few years, however, industrial expansion has begun to look less like healthy growth and more like a malignancy, and when status quo apologists argue that our economy can't tolerate anything less than a certain percentage of industrial bloating each year, they do nothing to lessen our feeling that the economy is in the grip of some runaway disease. People say that surplus wealth encourages research and development, but usually this means finding some new product to foist on the public rather than meeting our current pressing needs. Putting our fate in the hands of corporations, in other words, has failed. Perhaps they should be discouraged, while we reorient our national energies in some new direction. Left to themselves, corporations tend to go toward more-of-the-same-only-bigger, and that isn't quite what we need right now. The educated are showing an increasing hunger for balance and quality in their lives, while the poor are simply showing increasing hunger. Corporations are meeting neither of these needs, and we can't use any more widgets. What we want is to live decent lives, and for every move American corporations make to bring this about, they make two more to prevent it.

This is not an argument for adding more power and wealth to our bulging federal bureaucracy. At the moment I'm only concerned with changing our ways of thinking about the economy— instead of just manipulating economic mechanisms, I want to start matching our needs and desires to our energies and resources. If we were to start rewarding efforts to create livable human environments instead of giving prizes for professional rapacity, many of the economic institutions we tinker with so much might betray a surprising resiliency and adapt themselves to our new wishes.

How We're Doing

We're an energetic and highly skilled people and don't lack for tools and resources. If there must be increasing unemployment when it comes to producing and marketing the unnecessary, the same can't be said about meeting real needs. Must a powerful, wealthy nation keep people monotonously doing meaningless tasks in order to avoid starving to death, when those same people have severe needs of their own that go neglected? We seem unwilling to recognize and deal with the fact that our economy rests on a profound misdirection of energy. This timidity is catastrophic. All the economic BandAids in the world can't hide the fact that we've been spending our resources and labor stupidly for decades, and are beginning to pay the price.

Money, after all, is symbolic, not real—an illusion that we agree to share for convenience. We've been staring at it for so long, listening to the hypnotic droning of professional economists, that we've forgotten it's just a mechanism for using our energies to meet our needs. Money is a way of matching needs and resources, but when we get caught up in it, and treat it as having value in its own right, it fails to perform even this rudimentary function.

One way to put all this in perspective is to compare few vital statistics of modern and primitive societies. Most hunter-gatherer societies today live under harsh environmental conditions, on land areas that are unattractive to their more affluent agricultural neighbors. Yet in most ways their lives are more pleasant and secure than those of peasants or workers in "wealthy" societies. Contrary to capitalist mythology, they don't even work very hard—usually about twelve to twenty hours a week.[4]

Our ideas about the miseries of "underdeveloped" life come largely from agricultural societies in Africa, Asia, and Latin America in which peasants eke out a dubious existence under conditions of profound class exploitation. Any society with class divisions tends to be very hard on the poor, who must support not only themselves but the rich as well, and the rich have big stomachs. This is one of the ironies of welfare—that people seldom complain about having to support the rich but get quite incensed at having to support the poor, even though their wants are more modest. The idea, I suppose, is that the rich are quite enough of a welfare burden as it is, without the poor adding themselves to it.

In any case, modern fantasies of primitive life would have us believe that people in such societies slave from dawn to dusk, that only the fit survive, that children must work at an early age, that those who don't work, don't eat. The truth is that although most huntergatherer tribes work less than half the hours we do, many manage to support up to 40 percent of their population that doesn't work at all. Among the Bushmen, for example, men don't work regularly until they marry, between the ages of twenty and twenty-five. Modernist propaganda would have it that only a highly developed industrial society can support a large nonprod uctive population, but in fact we do barely as well as the hunter-gatherers, although we work a great deal harder and have many more resources.[5]

The success of such societies comes from their having opted out of the struggle for rich arable land and other scarce resources—deciding to content themselves with consuming what's available rather than trying to create surpluses. They have few possessions, for the insatiable demand for material goods that we think is innately human is in fact a symptom of decay that afflicts people when their communities have been disrupted and fractured.[6]

I'm not suggesting that we should all hurl ourselves back to Nature, even if it were possible and people were so inclined. I only want to expose misleading advertising. Modern civilization has brought us a huge assortment of goods but very little else. For every disease it has wiped out, it has created another through chronic stress. Current death rates for men of seventy are higher than they were in the middle of the last century. Death rates in the United States are still lowest among rural people who have the least access to medical facilities, while

Puerto Rico has lower death rates than the United States as a whole for all ages over twenty-five. People in modern industrial societies work harder than under any previous economic condition, with the exception of some of the most severe forms of agricultural exploitation. No one can claim that people are happier under modern conditions, or more refined, or less bloodthirsty; for every stress indicator—crime, mental illness, suicide, chronic disease, and so forth—increases with the spread of civilization.[7]

This is not to say it was all a mistake. We traveled this route and our history is a part of us; from this spot we can move in any number of directions. There's no particular point in saying every step we took was either a brilliant choice or a stupid blunder. Clearly there have been plenty of both. I want to focus on the mistakes not to discredit our past but to help us make sensible choices now. Sometimes people are better off reversing a bad step than they would be if they had never blundered in the first place. I'd rather have our mistakes in the past than in the future, for a blunder recognized is an inoculation against future perils. The important thing is what we do now with the knowledge we have about what our ancient choices have cost us.

Property and Ecology

One of the most costly of our ancient blunders was creating the idea of property. Usually when acquiring a new piece of property, we try to make sure it wasn't stolen, but this is a strange scruple since originally all property was stolen. The land and its products belonged to anyone and everyone to use, until forcibly expropriated by those powerful enough to do so. Most American Indians didn't feel they "owned" the land, for example—they shared its use with all living beings until they were forced off it. A title search regarding a piece of real estate just establishes that it wasn't *recently* stolen.

Social fictions have their uses, and pretending that a person could possess a piece of the earth, a slice of river, or a lump of ocean was an interesting intellectual excursion. But no solution of our ecological problems is compatible with the idea of property held in perpetuity. If I own a piece of land or a factory—if

it can be considered an extension of my ego—then it's a vital
democratic right that I be able to do with it what I want. Since
it's an extension of me, it is, like my body, mine to dispose of.
Anything else would be slavery. Although my use of it may on
occasion be restricted, the burden of proof is on others to establish my social and ecological responsibility. I can put my own soot
into my own section of air, my own poison in my slice of river,
and so on. The interdependence between my part of the air or
water and everyone else's is obscured by the idea of property.

The most obvious way property muddles our thought is by
creating artificial boundaries. It teaches us to draw arbitrary lines
that don't correspond to any natural relationships—lines that obscure and confound our understanding of nature. Imagine what
would happen if we were to proceed in the same way in physi-
ology: "the hand's territory continues up to this line, and from
there on it is the wrist's; the hand owns this part of this artery,
nerve, vein, muscle, the wrist owns that part." We would never
understand anything about the way the body functions, just as
we now understand virtually nothing about the functioning of our
environment taken as a whole.

Our ideas about our relationship to the environment are like
the "glove paralyses" of nineteenth century conversion hysterics.
Physicians knew these paralyses were psychological because the
hand isn't an anatomical unit—it can't be physically paralyzed
without affecting nerves in the arm, and so on. In the same way,
a piece of real estate isn't a meaningful unit in any natural system. Property tends, in fact, to cloud our awareness of such systems. Imagine slicing an organism into pieces, giving each piece
a separate identity, and then asking them all to organize themselves into a whole again—assuming the pieces could be kept
alive, with their slices of nerve, artery, muscle, and bone, somehow intact. Think of the arguments: the largest pieces claiming
superiority and demanding to be on top or at the center, tiny
slices of brain being shoved to the periphery because of bad
color. Frankenstein himself could do no worse at creating an organism.

We use land, water, air, objects, people. To own people is
considered barbaric—civilized people rent one another's ser-

vices. But to claim permanent possession of a piece of the environment is also a kind of enslavement. The environment ultimately belongs to everyone, for we all share its use, all depend on it for survival, and must all cooperate in maintaining it to ensure that survival. No one can reasonably claim a permanent right to a piece of land or water or sky. In the first place, no one lives permanently.

The property habit, however, is basic to our thinking. We're always cutting things into little pieces, sorting the pieces into compartments, and then wondering why we can't understand how anything works as a whole. Maybe that's why more people know how to divide an animal into cuts of meat than know how it works physiologically. Modern medicine itself has never recovered from the fact that it was founded on cutting up corpses: for since people tend to dissociate themselves from the dead, this gave medical scientists the unfortunate impression that they were somehow separate from what they studied. They never learned to relate their own feelings to what they were doing—to treat life as something they themselves were part of. Bodies were just pieces of meat, pieces of property. As scientists they owned them, but owed them nothing.

How you view the world affects how you treat it and hence how it reacts. If you see your environment as a collection of separately-owned pieces of dead meat, it soon will be. But since our whole culture is based on property, how can we possibly change?

We might at least begin to recognize that nothing is forever—that one can lease and use but never really possess. Even our bodies are returned to the earth at the end of our term, and to treat a less intimate piece of property as a permanent possession seems the height of folly. We can begin changing our thinking about property by realizing that ownership is just a grandiose way of talking about a long lease. If property is anti-ecological, inheriting it is even more so.

Is Change Inevitable?

I've said several times that to change our lives we need both to change the way we think about the world and to change those parts of the world that help make us think that way. Much of what

I write is a record of my own efforts to rethink my view of the world—written in the hope that it will be of use to others making the same attempt. This chapter is something of a departure in that it calls attention to needed changes in institutions that affect the way we think. I feel less comfortable suggesting these kinds of changes since I haven't personally experienced them, and since I'm somewhat mistrustful of change legislated from a distance. I'd like to digress a little here to say why I feel obliged to do so.

In my book *Earthwalk* I emphasize the fact that change is an ongoing process that we participate in, but never really "initiate." I discuss the process I call "social eversion," by which things turn into their opposite when pushed to an extreme. These three examples of social eversion will illustrate what I mean:

1) Pop futurologists look at our burgeoning technology and predict more of the same, unendingly. But nature is more imaginative. Technological advance has its own self-destruct mechanism built into it. For as change becomes more rapid, reaching the point of almost instant obsolescence, it becomes more and more impossible to plan for the future. But the whole structure of technological society is built on planning, looking ahead, living in the future, postponing gratification in favor of attending to tasks, and so on. So as technology pushes into the future, it also *undermines* the future, forcing people to turn to the present for security and gratification. People look to the here-and-now, to bodily pleasures, to nature and the spirit, and to traditions—all things that disarm and enfeeble the technological impulse. This has already begun to happen, arousing the fury of conservatives like Alvin Toffler, who see it as a betrayal of modernity, not realizing that in fact it's a part of the process of change itself.

2) A trivial but therefore rather clear-cut example of eversion is the development of acronyms. As our society became more bureaucratized, the names of organizations became more complex and technical. Names like Kellogg or Campbells became less frequent, names like International Business Machines and General Dynamics more frequent. Meanwhile, government agencies proliferated, starting with the New Deal, and people began using initials to abbreviate complex titles: NRA, CCC,

WPA, NLRB, FBI, and so on. This was extended to private organizations as well, from CIO, GE, and AFL to (later) SNCC and SDS. But by the fifties a third trend was in process—the creation of acronyms. Initials were combined to form a word, like CORE, ACTION, MERIT, or VISTA. At first this was fortuitous—if the initials vaguely sounded like a word, that word (NATO and RAND are examples) began to be used in place of the initials. The word didn't have to be a real one at first—many corporations even before the thirties had names that were acronyms: Socony, Nabisco, Esso, and so on. But now the effort was made to find a formal, official-sounding title whose initials would form an existing word—a word with connotations that expressed something positive and meaningful about the organization. In other words, the acronym became more important than the title whose initials composed it. By now the flight from technical pomposity has brought us almost full circle, from names to titles to initials to acronyms and back to names. Once the environment is filled with evocative words like MORAL, AIM, NOW, ROAR, and ACID, it's only a matter of time before people begin to abandon officialsounding titles altogether and return to simple names, which is already the case for many alternative institutions. This example is useful because it makes the eversion process so clear: titles don't just keep getting longer and drearier— the very length *creates* the acronym. In other words, excess creates its own antidote so long as the rest of the system (human beings, in this case) retains some remnant of balance and harmony.

3) A final example has to do with the relation of the household to the market economy, described in Scott Burns's *Home, Inc.* The history of industrial capitalism is one long story of economic functions being sucked out of the home and into the industrial market, where they come to be performed by larger and larger organizations. This leaves individuals more and more atomized—tied together less and less by shared economic activities at the local community level—and increasingly at the mercy of huge and distant bureaucracies, whose economic advantages are progressively undermined by distribution costs. In response, people withdraw more and more into the household, to the point

where, as Burns shows, the household economy, in both labor and capital assets, has begun to outstrip the market economy and threatens to make money itself obsolete.[8]

* * *

All these examples suggest a kind of natural organic rhythm in social change, as if human social experiments were somehow in balance, but over such a long time span that we can't perceive it. They imply self-corrective mechanisms that are actually working. Not that change is completely circular—when history loops back, it never returns to the same place. But neither does it ever keep going indefinitely in the same direction.

The idea of social eversion might suggest to some people that things will work themselves out anyway, whether we do anything or not. This thought pleases some people and annoys others. But all arguments about fatalism rest on a false individualistic assumption: that our actions or inactions are somehow *outside* a process like social eversion—that we can choose to participate or not participate. But whatever we do or *don't do* is a response to some part of that process, and continues it in some direction.

FREEING BLOCKED SOCIAL ENERGY

What *I'm* responding to is a sense of frustration that so many healthy new growths in our society are at some point blocked by the overwhelming force and rigidity of economic inequality. There are a great many things changing in our society—basic attitude changes of a healthful kind are diffusing from the fringes to the center, or, if you like, from the bottom up. But there's a roof that always blocks further healthy growth—a ceiling of concentrated economic power that holds us back, frustrates change, locks in flexibility. It has long since used up whatever potential it had for doing anything creative—all its energies are devoted to preserving its special advantaged position. It buys protection and favors from Presidents; cheats, robs, and poisons the consumer with impunity; continually widens the gap between rich and poor; and ties up in its own greed the resources we need to solve the problems of our society.

So long as that roof exists, all healthy social growths will be stunted. Each will grow to the same point and then get stuck. My emphasis on economic issues in this chapter is an effort to remove a rigid obstacle so that natural curative processes can take their course. As I suggested earlier, there is great vitality in our society but it's constantly hamstrung by economic inequality.

One source of rigidity is the fact that the wealthy and educated can so often buy out of social problems, leaving the poor to choose between short-run suffering and long-run suffering. Affluent liberals ean afford to favor things like forced busing and restrictions on stripmining—issues they understand but are unaffected by. They can see that it may be better in the long run if poor people undergo some immediate discomfort or suffering. But the poor don't have the luxury of longrun thinking, while those who have the resources to cope with such problems aren't available. While the poor are being racist or resisting environmental controls, the well-to-do are off at the beach.

The rich, of course, are also racist, and responsible for most pollution. What I'm saying is that either way they can buy out: from paying the price if they act in bad faith, from having to confront the issue if they're well-intentioned. This is not to imply that the "larger view" of affluent liberals is necessarily more correct than the impulsive reactions of the poor. My point is that *both reactions suffer—the one from having too little leisure to appraise the problem, the other from having too much.* It's easy to be against strip-mining if you're not a miner—it means you never have to face up to the real quesrions about changing our work priorities, redirecting our national energies. A liberal legislator, for example, may fight hard for an environmental decision that eliminates jobs temporarily, on the assumption that unemployment benefits exist to handle such transitions. But then he compromises with conservative colleagues (now that he's won *his* point) who want to cut unemployment and welfare benefits. And when all that's done, he's too tired to think of changing the direction of the society, so he goes on vacation. Meanwhile nothing forces him to change the direction or style of his *own* life, which helps keep the society firmly on its old track.

Inequality, then, is a severe obstacle to the natural flow of healing social energy. In the past, vast accumulations of wealth were justified as providing the wherewithal to initiate grandiose projects, research new potential products, and pay people enough money to get them to work on these projects rather than on the things they'd enjoy working on. But times have changed, and what we now need is not more vastness or new products but a lot of small changes.

Mechanisms for removing the worst inequalities are available and have often been discussed. Until now, however, the wealthy have been able to buy off any severe challenge to their special position. So long as ordinary people could imagine that they themselves might someday be rich, it was difficult to mobilize popular sentiment for equality. But as these pipedreams dissipate, the prospects for equalizing wealth begin to brighten.

Now talk of equality arouses irritation among the rich and educated, who would like to believe that their favored positions are somehow deserved. Equality sounds monotonous to them—they flutter at the thought of a society composed solely of Levittowns filled with beer-drinking television viewers who haven't read Céline, can't pronounce Pouilly-Fuissé, and haven't the faintest idea what "ambience" means.

Some years ago a Harvard psychologist, Richard Herrnstein, suggested that inequality of wealth, status, and opportunity was a blessing in disguise, since if all these were stripped away, inequality would be determined by innate differences in ability from which there would be no appeal.[9] For Herrnstein, ability or competence was something I.Q. tests measure—verbal facility and logic. He thought that with equal opportunity, people with high I.Q.'s would oppress everyone else with perfect justification, since they were of most value to society. The flaw in this argument is that people considered especially bright have managed to make an unusual number of disastrously short-sighted decisions during the past fifty years, and I'm far from convinced that we need still more of what they have to offer. If the opposite of what they've been giving us is stupidity, then I would maintain that stupidity is a priceless national resource.

"Intelligence" tests measure your ability to achieve personal success in our society at all costs. They tap your willingness to ignore your body, your physical surroundings, your human relationships—tyour entire environment, in fact—in a blind, single-minded endeavor to convince the tester that you're smart. This may *be* a kind of smartness, but it hardly seems like wisdom. The fact that we value it so highly helps explain the trouble we're in.

Our society needs people with other gifts—people less narrow, less singleminded, less slavish, less self-serving—people who would "waste time" trying to relate to the tester as a person, or listening to their stomachs rumble, or thinking how cold and ugly the building was. We need people who are more "distractable"—interested in the process as well as the product. Furthermore, our society contains such people, even if the Herrnsteins can't see them. Equality would make their gifts available to us, and provide a balance our society now lacks.

Inheritance

One of the easiest ways of equalizing wealth would be simply to eliminate all forms of inheritance. It has the great advantage that no one can say money is being taken from the industrious and given to the idle, since that's what inheritance does anyway. The change at the very worst would be one of taking from the idle rich and giving to the idle poor. For when you get right down to it, inheritance is just an elaborate and expensive welfare program for rich people.

Some might object that people need to provide for their children, particularly for their education. This is reasonable, but perhaps the term "provide" need not extend to a million-dollar trust fund. What I'm seeking here is a small beginning toward giving everyone in the United States the same start. At present "equality of opportunity" is a horrendously bad joke in America. While some people can't even feed their children, others are able to extend their mammoth narcissism unto the third and fourth generation. It would be very hard to make a case that inherited wealth ever did *anyone* any good. It flatters parental egos to be able to enrich and enfeeble their children, but this dubious benefit is vastly outweighed by the many more profitable uses to which the money could be put.

If we truly believe, for example, in equal educational opportunity, it would be more appropriate to transfer such money to scholarship funds for *all* bereaved children, rich or poor. It could also pay for medical and other necessary expenses.

There are two practical problems about such a plan. The first is legal evasion, such as we now have in relation to inheritance taxes: the rich will always find loopholes if we're not vigilant. But trust funds can be made illegal, gifts can be limited in size, and so on. The more stringent the laws, the more money will pour into the scholarship funds, and the more successfully these funds operate, the weaker the motivation to cheat them. As for enforcement, it essentially pays for itself.

The second problem is the old one about increasing the federal bureaucracy. To avoid this, one could create local foundations whose task would be simply to receive and disburse such funds, under public scrutiny and in accord with federal regulations. We already have a substantial government apparatus for regulating trust funds, collecting inheritance taxes, and overseeing private foundations, as well as an enormous private legal structure for evading regulations as much as possible within the law.

Difficulties will always arise in such plans as long as we're taught to think individualistically. If we ignore the fact that our destinies are linked together, evading laws that try to advance the welfare of the whole community will always be considered fair game. What these suggestions point toward is the creation of benign local environments in which people can lead safe, satisfying, and stimulating lives. To bequeath such an environment to a child is far better than just leaving her an expensive escape ticket. Trying to buy those escape tickets is what messes up the environment in the first place.

Taxes

At present our tax methods are often regressive: sales taxes, state lotteries—even income taxes, given the loopholes available—fall more heavily on lower income people. But income taxes provide the best opportunity for equalizing wealth and it could be done, as I said in the last chapter, by simply eliminating all deductions, exemptions, and allowances. Most of these favor the wealthy, who are thus enriched at public expense. Furthermore, tax money is used to do the things we as a nation have decided we want. To give tax relief as an incentive is to deny the common benefits taxes bring.

We now give a tax break to the hustler who bribes a public or private official with expensive lunches and junkets—subsidizing fancy restaurants and hotels. The wealthiest homeowners get the biggest tax breaks for mortgage interest, while poor renters get nothing at all. And the loss to the government is greater from the rich: a $100 deduction is worth $70 to someone in the highest bracket, only $14 to someone in the lowest. Many provisions benefit the wealthy almost exclusively: reduction of taxes on capital gains, on interest earned from certain bonds, on dividends; allowances for depreciation; deductions for charitable contributions; and so on.

In a society that touts the value of work, it seems odd that unearned income isn't taxed more severely. This omission is always justified with stories of aged widows living on income from securities and suffering from chronic inflation. But there are two things wrong with this picture. First, there are many more aged widows struggling along *without* such income; it would make a lot more sense to give direct aid to the elderly. Second, most unearned income accrues to the very wealthy. If income taxes are properly progressive, the effect on the poor of taxing unearned income fairly would be minor indeed.

A more serious rationale for subsidizing the wealthy is the fear of discouraging investment. This is the fundamental argument rich people trot out when any attempt made to get them to pay their fair share for community needs: people won't invest in corporations, which will cut back production, causing unemployment and large-scale bankruptcy. Investments are needed, it's argued, so that enough capital will be available for plant renewal and expansion, and so on.

The argument sounds reasonable but it contains several flaws. In the first place, the individual investor plays a relatively minor role today in industrial growth. Corporations themselves, along with pension funds and other large investment trusts, are the major source. The rich individual is no longer essential to the investment market. A second problem with tying investment too closely to greed is the irrelevant motives money always introduces. If you attract investment by the amount of income it yields, for example, then corporations are obviously motivated to pad dividends at the expense of all the wonderful things they're supposed to be doing with those investments. Ma Bell, which for decades has had the reputation of paying good dividends, has in recent years had the effrontery to demand rate increases to improve their deteriorating service. In other words, they demanded a reward for doing a poor job, and forced poor people to pick up the tab, paying higher rates for poor service so that rich investors could continue getting dividends.

The trouble with money is that it always distracts people from the goal of providing goods and services and creating a better environment. Somehow it always ends up just inflating itself. Our economy is based on spending billions to persuade people that happiness is buying things, and then insisting that the only way to have a viable economy is to make things for people to buy so they'll have jobs and get enough money to buy things. Now, if you can base a whole economy on such a silly circularity, you could just as well base it on some other silly circularity. How about paying people to do what needs to be done so that they'll have enough money to pay for the financing of what needs to be done?

Supposing a corporation goes bankrupt. We are now in a position to decide whether it's doing something necessary for us or not. If so, it would make more sense to reconstitute the corporation on a nonprofit basis. If not, we're better off without it, and its employees would be better employed doing something else. This isn't to be treated lightly, of course, but it isn't to be avoided either. Our society is in the process of retooling itself—of spending its energies in less frivolous and self-defeating ways. Paying people to do what isn't of any value is no solution, even though it may seem to ward off trouble for the time being. Reemployment is an epidemic problem in our society and has to be dealt with directly. *There's probably no more important task ahead of us than finding a way for people to make a living being useful to the community.*

The "don't discourage investment" argument is also used against the excess profits tax. But what are excess profits actually used for? To some extent they're used by the corporation to gobble up other companies and expand, rather than to do its job better. We need, over the next decades, to wean corporations away from the idea that their goal is to enrich stockholders rather than to fulfill consumer needs. An excess profits tax does precisely what it implies: it discourages a corporation from sacrificing everything to short-run monetary gains. A strong excess profits tax, for example, might have softened the 1973–74 oil crisis considerably.

What all these arguments boil down to is that taxing the rich discourages people from putting all their energy into trying to become rich—an outcome devoutly to be wished at this stage of our history.

If, for example, we merely eliminated all tax loopholes for the richest one-seventh of the population, *we would have enough money to create four million jobs paying $7500 a year.*[10] Millions now collecting unemployment insurance could be hired as policemen, firemen, teachers, sanitation workers; or fill in newly-created jobs—working on environmental problems, pollution control, and so on. It becomes ridiculous to talk of the high cost of welfare when these options are open to us.

Another step would be a 100 percent tax on all income over $100,000 a year. Such a sum scarcely discourages enterprise, and anything more seems excessively greedy. We can no longer afford to create and support millionaires. The motivation to make millions hasn't led to any social benefits for some time (if it ever did). Opportunities for acquiring such wealth are almost always damaging to the society and the economy. The formation of conglomerates, for example, that feed like vampires on healthy corporations, draining them not only of wealth, but of quality and morale—of what possible social benefit are they? In the last analysis the pursuit of wealth creates nothing but the pursuit of wealth.

Size

After some point, every increase in size is accompanied by a decrease in quality. While the "efficiencies" of size are usually illusory, as I pointed out before, the dangers of size are not. Concentrations of wealth and power are an ever-present danger to the democratic process, as people have been saying for the past century. Above a certain size a corporation can't even fail, since the government will keep it alive with transfusions of money, much the way doctors mechanically prolong dying. Once bloated beyond some point, a corporation is virtually immortal, and can do whatever violence it likes to its portion of the earth's anatomy before it will be allowed to fail and eliminate jobs. Lockheed and Penn Central are our most extreme cases of welfare fraud. The anti-trust approach to this problem hasn't been notably success-ful, and perhaps any attempt to attack the issue directly is futile but I would like to suggest an alternative.

The size of an overall budget determines an organization's ca-pacity for lobbying, advertising, bribing, and generally exercis-ing political clout by legal or illegal means. One way of encour-aging the formation of smaller and more responsive corporate units would be to impose a size surtax. Any corporation with assets or income over a certain figure would be forced to pay a steeply progressive surtax. Furthermore, if it held more than a certain percentage (ten percent, say) of the shares of another cor-poration, the total assets or income of that corporation would be added to its own in calculating the tax bill. Such a plan would not only be a rich source of tax revenue, it would also discourage one of the most pathological trends in corporate America today, trans-fusing back into the community the energy sucked from it by ambitious psychopaths.

Where we *need* size, it will force itself upon us. Usually we turn to it out of sloth. A competitive system leads automatically to corporate acromegaly, since somebody has to win: if economic life is a tournament, it follows logically that there will be only one corporation left at the latter day. Until we achieve a more cooperative mentality, smallness needs to be given direct sup-port.

Guaranteed Survival, Conditional Wealth

The idea of a guaranteed national income has been frequently advanced in recent years, and is usually regarded as the ultimate in liberal socialism: the welfare state taking care of everyone. Now there's no advantage in trying to limit the size of private

1682

CHAPTER 7

corporations if this simply leads to a greater concentration of power in Washington. But community needs and obligations can be handled in many ways, and the tendency to dump them into the lap of the federal government is just inertia. Most things that can be nationalized can be localized as well.

The idea of a guaranteed national income is of interest here for two reasons. First, it is a way of limiting inequality of wealth: having put on a ceiling, it seems appropriate to put in a floor, just as we now guarantee the survival of the rich, we would begin to guarantee that of the poor. Second, and more important, it's a means of shifting the way we allocate human energy: *we would no longer have to keep people busy doing useless and destructive things in order to keep unemployment down.* In a sense, it would provide a cushion while we redistribute our national energies. This redistribution seems so vital to me as to outweigh any drawbacks.

The principle underlying the guaranteed income is that no one should be allowed, in so wealthy a society, to die of starvation, exposure, or neglect, no matter how "undeserving" he or she might be that any society worth its salt should be able to feed, clothe, and shelter its members. Otherwise, what good is it? Most "primitive" societies do as much.

As many have pointed out, we now have a welter of separate programs—welfare, food stamps, unemployment benefits, social security—so the cost of a guaranteed income plan would not be as great as it might seem. Everyone would receive a small fixed income—increasing presumably with advanced age—enough for food, necessary clothing, and shelter. Survival would then be separated from labor. People would work because they enjoyed it, or to obtain desired goods and services but not out of desperation. Many people would work part-time, or in volunteer work, doing things that need doing, that money doesn't buy today. In other words, it would help bring about a closer match between what people do and what needs doing. It would help eliminate the absurdities we find around us today, such as laying off government employees to save money and then having to pay them welfare and unemployment benefits. It would also remove another problem: we would no longer be subject to economic disaster if people did their jobs well.

These proposals have been discussed not because they're important in themselves, but because they help focus our attention on the question of how we use our human and physical resources. They're *preliminary* to more basic changes that are needed, and are necessary only to avoid perverting and warping those changes.

The basic changes in our society will be gradual and spontaneous: changes in motives and values, changes in sex roles, the withdrawal of energy from the market economy, and the politicizing of the corporation. The last is a particularly important trend, since direct political confrontations with corporate policy—in stockholder's meetings, through boycotts, picketing, demonstrations, strikes, class-action suits, or whatever—break down one of individualism's most treasured principles: the idea that a corporation has no responsibility to the community so long as it breaks no laws. The politicizing of the corporation is one of the most important ways Americans are beginning to recognize their interconnectedness.[ii]

Money and Costs

Every field is a mixture of knowledge and hocus-pocus, from the most rigorous science down to medicine, psychology, the social sciences, and astrology. But it would be hard to find a field more fraught with mystification than economics, partly because it figures so strongly in political disputes. Most people feel a vague but insistent skepticism about professional economists, the certainty of whose predictive pronouncements varies inversely with their accuracy. Government economists tend to take the position that the public is just naïve, ignorant of the complexities of economic processes. The fact that their own superior knowledge rarely leads them to agree with one another doesn't seem to distress them. Nor does the fact that our economy has become progressively sicker in response to their ministrations.

Frankly, I have more faith in the intuitive good sense of the most ignorant householder than I do in the convoluted reasoning of a learned expert whose eyes are bedazzled by money. Economists tend, by the nature of their discipline, to make money an end rather than a means. They seem unable to think beyond it—beyond what will happen to investments, markets, savings, income, and so on. It's as if we asked a man how to improve communication between people and he gave us a long lecture on how to take good care of telephone wire.

Money is simply a way of making it possible for us all not to have to work at the same tasks. The serious questions of our day have to do with our uncertainties about what we want to work on—how we want to spend our energies. Some people have gone off to the country to live almost self-sufficiently—shutting themselves off from the money economy as much as possible. Most of us, however, want more than such a life permits. But working at a job you don't like, making something you don't want, for people who will have to be persuaded to buy it, is a sign that something is amiss—something that can't be justified in terms of the "needs of the economy." The money economy, after all, is supposed to serve our needs; yet here we are, breaking our necks to feed something that no longer even works for us. When government economists speak, it isn't just that the cart always comes before the horse—we never see the horse at all.

One source of mystification is the idea of cost. Things "cost money" only from the buyer's short-term viewpoint. When you move to the scale of national economics, nothing costs anything in money terms, since money is just a medium of exchange—it just moves around among us. What a government spends to create employment comes back in taxes, and so on. At the societal level it makes sense to talk about cost not in money terms, but in terms of labor and resources. Any task we undertake as a nation—whether it's feeding the poor, encouraging people to get rich, sending people to the moon, or fighting wars—costs us labor and material resources. We have to decide what kinds of things are worth spending our own energy and resources on, not what "money" they cost.

This way of thinking clears up a lot of things. We realize that war doesn't "stimulate the economy" so much as it cons us into working harder and exhausting our natural resources. When we use labor and materials to produce hardware that's dumped on foreign countries and destroyed, it's a total loss. Nothing comes back—no taxes, no benefits, no goodwill, no nothing. We also realize that tempting people to swindle one another encourages labor that does nothing for the community. Finally, we realize that working to help each other is likely to increase the labor pool and create a benign environment.

The question is not, how do we spend our money, but how do we spend our time and energy? If we spend it in competition, we get little back. If we spend it in cooperation, we get it all back.

We attribute what prosperity we have to competition because we call our economy "competitive." It's a matter of bookkeeping. What we actually produce in goods and services is done largely through cooperative labor. The *pricing* is done more or less competitively (although not nearly as much as it's supposed to be). Since we like to measure things in dollar terms, we credit competition with all that we achieve, even though the actual labor was cooperative. There's no dollar valuation on what's cooperatively or lovingly created. As Scott Burns points out, however, cooperative household labor alone is the biggest industry we have. Whereas money is needed to force people to work competitively at unrewarding labor, cooperative work often doesn't even require a money reward, yet its products are superior.

If competition were all there was to America, we wouldn't be here now. Cooperation brought about our independence from England and the formation of our own Constitution. The most vital parts of our heritage are cooperative, not competitive— they represent people working together in recognition of a common suffering, common needs, and a common destiny—economic, political, and personal. We've always been in the same boat—we've just been too busy staring over the side to recognize it.

Looking Ahead

Suppose we begin at a natural beginning, with what we want as a people. Obviously as one person I can only guess at this. But for purposes of argument, let's say that we all want some combination of the following:

1) Safety—freedom from local or national violence;
2) Survival—food, clothing, shelter, and access to health care;
3) Beautiful and healthy surroundings;

4) Freedom to pursue our own unique personal goals as best we can—whether this means pleasure, prestige, wealth, emotional growth, learning, moral development, or whatever.

Now it's easy to see that the fourth item is the tricky one, in that it isn't awfully compatible with the other three. Any form of authoritarian socialism could easily provide a nation as wealthy as ours with the first three blessings. On the other hand, our society as it is severely frustrates all three in order to provide the fourth.

The fourth goal is usually phrased in a way that assumes a competitive world but in fact competition isn't essential to it. Nor do the first three require cooperation—mere conformity to a benign despot would do the trick.

Cooperation is desperately needed, however, in any attempt to unite all of these contrary goals in one society. As a people we're too used to following our own initiative to give it up—that freedom is embedded in the tissues of our bodies. Yet the price we've paid to make it our ruling principle seems more and more intolerable to more and more people.

When there are enough cars, everyone has her own. When there's a shortage, people form car pools. While shortages usually make people behave more competitively, they can also force us to broaden our view of a situation. For example, at the height of the last gasoline shortage, it could not be said that there was any shortage of transportation facilities. What was in short supply was merely the ability to be *alone* while moving from one place to another.

The world is full of bounties—full of energy, matter, information, gratification. Shortages can only be local or temporary. This means that every time we broaden our vision, either in time or space, we define scarcity out of existence. In the Bible, Joseph did this when he foresaw a fourteen-year cycle of crop production. A recognition of world-wide interdependence would have the same impact today.

We experience shortages when our vision is too narrow—when we think of cars rather than transportation, steel rather than metal, oil rather than energy, paper rather than information-storage-and-distribution, and so on. And especially when we fail to see interconnections: thus more cars create a "shortage" of roads, and more roads create a "shortage" of land, trees, grass, air, and so on.

Our most profound mental block as a people is our inability to think of things in relation to each other—our insistence on looking at only one thing at a time. We always think that getting more of something will make us happy, and that a *lot* more will make us happier still. We have a hard time understanding that health, or happiness, or true prosperity is achieved when things are in balance. We often fail to notice that shortages and other crises seem to raise the general level of cheerfulness in the society. For any shortage has to be looked at in relation to all others—a shortage of gas alleviates 1) the shortage of exercise, 2) the shortage of life (from accidents), 3) the shortage of human contact, and so on.

Working together to create more balance can remove our sense of scarcity. We have abundance of wealth but it is badly distributed. We have abundance of time but it is experienced in mental absentia. We have a surplus of information but it is completely uncoordinated. We have no end of real gratification available to us, but we run in all directions looking for fantasied ones.

For many people, talking about coordination and balance means "planning"—some form of government intervention, from mild regulation to complete government ownership. But planners just seem to make the same mistakes on a larger scale— always maximizing or minimizing something, never optimizing it in relation to something else. The Soviet Union's economic disasters have been in keeping with its size, and no one could say that our own experiments in government intervention have been felicitous. Who could deliberately invent anything more asinine than the farm program born in the New Deal, in which farmers were simultaneously paid to grow more and to grow less? What could be better designed to make farming a big business rather than a family enterprise? Or to encourage the overuse of chemical fertilizers and insecticides?

On the contrary, the slight amount of anarchy left in the business community, after years of increasing monopoly, corporate obesity, and governmental liaisons, is what gives it the only flexibility and environmental sensitivity it has. If we can make size unprofitable, grotesque wealth unattainable, and survival indisputable, then free enterprise would no longer be oppressive, since what happened to the market economy would no longer have life-or-death significance to the society as a whole.

In *Communitas,* written in 1947, Percival and Paul Goodman suggested that subsistence should be withdrawn from the market economy. Scott Burns expands on this notion, observing that since the subsistence economy could be supported by one-fourteenth of the work force, a person could earn a lifetime of food and shelter by working for three years. Another year would buy twenty years of education and so on.[12]

Money distracts us from the issue of what we want to gain from our labor. Burns's idea of a National Service Corps, in which one would work for two or three years to obtain lifelong support, is one way of detaching the right to survive from the competitive greed and inhumanity that dominates the market economy. *Given any system of this kind, in which survival was guaranteed, we could abandon all our frantic manipulations of the economy, allow it to stumble, stagger, and fall through whatever readjustments were necessary, and concentrate our energies on whatever tasks need doing, and/or are personally satisfying to do.* Some people would work for money, some for satisfaction, some for service.

Changing from a money-based to a work-based economy would bring about a tremendous upheaval, equal to that of the early thirties, but at present it seems doubtful that such an upheaval can be avoided in any case. Unlike past economic unheavals, however, it would affect the rich more profoundly than the poor. More yachts for sale, fewer bread lines.

Burns observes that changes of this kind would tend to make the job market freer and more competitive. If survival is no longer dependent on work, poor people will no longer be forced to do the most unpleasant jobs. Except for the National Service Corps, the job market would be entirely governed by supply and demand; attractive jobs would pay very little while unattractive ones would be lucrative.

All these changes seek the same end: *instead of money being a freely-responsive force that controls the expression of human* energies, *human energy would be a freely-responsive force that would control the distribution of money.* Money in modern society has long been an unmanageable tail, wagging the body of human effort in inelegant spasms. The assertion of our right as living beings to wag our own tail should be the first priority in any program of social change.

It isn't much, but it's all we've got.

Notes

CHAPTER ONE

1. S. A. Stouffer, *Communism, Conformity, and Civil Liberties* (New York: Wiley, 1966), p. 164.

2. Ezra F. Vogel and Suzanne H. Vogel, "Permissive Dependency in Japan," in *Comparative Perspectives on Marriage and the Family*, H. Kent Geiger, editor (Boston: Little, Brown, 1968), pp. 68–77.

3. David B. Wilson, "No slogans; let's all fast," *The Boston Globe*, September 14, 1974.

4. David Riesman, *Individualism Reconsidered* (Garden City, New York: Doubleday Anchor, 1954), p. 27. This is a principle for which nature has shown a fine disregard—evolution proceeds on a diametrically opposite principle.

5. Jay Haley, "The Family of the Schizophrenic: A Model System," in *The Psychosocial Interior of the Family*, G. Handel, editor (Chicago: Aldine, 1967), pp. 271–272.

CHAPTER TWO

1. Frank Harvey, *Air War—Vietnam* (New York: Bantam, 1967), pp. 15, 29–30, 67, 100, 104, 108–109, 115.

2. Ibid., pp. 2, 16, 55–56, 107, 116.

3. Ibid., pp. 108, 141, 146–147, 150, 154; Robert Crichton, "Our Air War," *New York Review of Books*, IX, January 4, 1968, p. 3.

4. Harvey, pp. 54–57, 82ff; Crichton, pp. 3–4.

5. Harvey, pp. 57, 91–92, 102–104, 115, 174–175; Crichton, pp. 3–4.

6. Harvey, pp. 39–40, 62–63, 106–107, 126–127; Crichton, p. 4.

7. Harvey, pp. 65, 70, 72, 105, 109, 111, 112, 138, 150, 152; Crichton, pp. 3–4.

8. Harvey, pp. 5–8, 11, 63, 115.

9. Martha Wolfenstein and Nathan Leites, *Movies: A Psychological Study* (Glencoe, Illinois: Free Press, 1950), pp. 106ff, 149–174.

10. Margaret Mead, *Sex and Temperament in Three Primitive Societies* (New York: Mentor, 1950), p. 21.

CHAPTER THREE

1. See W. G. Bennis and P. E. Slater, *The Temporary Society* (New York: Harper and Row, 1968), Chapter 2.

2. John L. Fischer and Ann Fischer, "The New Englanders of Orchard Town, U.S.A.," in *Six Cultures: Studies of Child Rearing*, Beatrice Whiting, editor (New York: Wiley, 1963), pp. 921–928.

3. Benjamin Spock, *Baby and Child Care* (New York: Pocket Books, 1968), p. 12. See also pp. xvi, 10–23.

4. Ibid., pp. 563–564.

5. P. E. Slater, *The Glory of Hera* (Boston: Beacon Press, 1968), pp. 450–451. See also Bennis and Slater, *The Temporary Society*, pp. 91–92.

6. I am indebted to Dori Appel Slater for this observation.

7. Slater, *The Glory of Hera*, Chapters 1, 14, 15. Chapter 14 contains the relevant research and references.

8. Ibid.

CHAPTER FOUR

1. R. Lynn, "Anxiety and Economic Growth," in *Nature*, 219, 1968, pp. 765–766; S. Freud, *Civilization and Its Discontents* (London: Hogarth, 1953), pp. 74, 92, 142–143.

2. J. D. Unwin, *Sex and Culture* (London: Oxford, 1934); G. P. Murdock, "The Regulation of Premarital Sex Behavior," in *Process and Pattern in Culture*, R. A. Manners, editor (Chicago: Aldine, 1964), 399–410; P. E. Slater, *Footholds* (New York: Dutton, in press), Chapter 10. See also H. Marcuse, *Eros and Civilization* (Boston: Beacon Press, 1955); N. O. Brown, *Life Against Death* (New York: Vintage, 1959).

3. Fred Cottrell, *Energy and Society* (New York: McGraw-Hill, 1955), p. 4.

4. A. R. Holmberg, *Nomads of the Long Bow* (Washington: U.S. Government Printing Office, 1950).

5. For a discussion of the ways in which a particularly extreme form of sexual scarcity may have begun, see P. E. Slater, *The Glory of Hera* (Boston: Beacon Press, 1968), Chapters 1 and 15.

6. Cf., eg., C. S. Ford and F. A. Beach, *Patterns of Sexual Behavior* (New York: Harper, 1951), p. 78.

7. Cf. Denis de Rougemont, *Love in the Western World* (Garden City, New York: Doubleday Anchor, 1957); J. C. Flugel, *The Psychoanalytic Study of the Family* (London: Hogarth, 1957); S. Freud, *Collected Papers*, Vol. IV (London: Hogarth, 1953), pp. 192–216; Slater, *Footholds*, Chapter 12.

8. See Slater, *Footholds*, Chapter 9; Cf. also Stuart, *Narcissus* (New York: MacMillan, 1955).

9. Erving Goffman, *Behavior in Public Places* (New York: Free Press, 1963), pp. 56–59.

10. E. Cleaver, *Soul on Ice* (New York: McGraw-Hill, 1968), pp. 202–203.

CHAPTER FIVE

1. R. B. Lee and I. DeVore, *Man the Hunter* (Chicago: Aldine, 1968), pp. 39–40, 54, 223, 244.
2. Lewis Mumford, "The Fallacy of Systems," *Saturday Review of Literature*, XXXII, October 1949.
3. Leon Festinger, *A Theory of Cognitive Dissonance* (Stanford, California: Stanford University Press, 1965).

CHAPTER SIX

1. H. Marcuse, *An Essay on Liberation* (Boston: Beacon Press, 1969), p. 18.
2. S. Freud, *Civilization and Its Discontents* (London: Hogarth, 1953), pp. 46–48.
3. Alvin Smith, *Boston Sunday Globe*, January 5, 1969.
4. These ideas were brought together in W. G. Bennis and P. E. Slater, *The Temporary Society* (New York: Harper and Row, 1968), Chapters 1–3.

CHAPTER SEVEN

1. Scott Burns, *Home, Inc.* (New York: Doubleday, 1975), pp.6–7, 17.
2. Ibid., pp. 122–123.
3. Robert LeKachman, "Toward Equality through Employment," *Social Policy*, September/October, 1974.
4. Lee and DeVore, *Man the Hunter*, pp. 37, 54; M. Sahlins, *Stone Age Economics* (Chicago: Aldine, 1972), pp. 1–39.
5. Ibid ; Lee and DeVore, p. 36. Eskimo societies form a major exception to this rule.
6. T. R. Gurr, *Why Men Rebel* (Princeton University Press, 1970), pp.93–113.
7. These data are based on an important but as yet unpublished paper by Joseph Eyer, Department of Biology, University of Pennsylvania, Philadelphia, entitled, "Stress-Related Mortality and Social Organization."
8. This is a highly oversimplified rendition of Burns's thesis, which embraces a complex series of interlocking eversion processes. See especially Chapters 3, 12, and 17 of *Home, Inc.*
9. Richard Herrnstein, "I.Q.," *Atlantic Monthly*, September 1971. Cited in Burns, p. 127.
10. This figure and others cited are based on data from a U.S. Treasury Department study for 1974. See *The Boston Globe*, May 27, 1975.

11. David Vogel, "The Politicization of the Corporation," *Social Policy,* May/June 1974.

12. Burns, pp. 89–91; Percival and Paul Goodman, *Communitas* (New York: Vintage Books, 1960), pp. 188–194.